A-Level Year 1 & AS

Biology

Exam Board: OCR A

AS Biology exams can seem pretty daunting unless you get plenty of practice at answering exam-style questions — and, oh, would you look at that — this fabulous CGP Workbook is full of them.

The whole AS Biology course (and Year 1 of A-Level) is covered, with a wide range of questions to really test your wits. Full answers are included, along with top tips to help you grab all the marks you can when the big day comes.

By the time you've worked your way through all this, the real exams should have no impact on your ventricular contraction rate...

CGP

A-Level revision? It has to be CGP!

Published by CGP

Editors:
Charlotte Burrows, Ellen Burton, Daniel Fielding, Christopher Lindle, Emily Sheraton, Hayley Thompson.

Contributors:
Mark Ellingham, Emily Lucas, Ciara McGlade, Bethan Parry, Megan Pollard.

Phylogenetic tree on page 69: Jacobo Reyes-Velasco, Daren C. Card, Audra L. Andrew, Kyle J. Shaney, Richard H. Adams, Drew R. Schield, Nicholas R. Casewell, Stephen P. Mackessy, Todd A. Castoe; Expression of Venom Gene Homologs in Diverse Python Tissues Suggests a New Model for the Evolution of Snake Venom, Molecular Biology and Evolution, Volume 32, Issue 1, 1 January 2015, Pages 173–183. By permission of Oxford University Press.

ISBN: 978 1 78294 917 6

With thanks to Janet Cruse-Sawyer and Rachael Rogers for the proofreading.

With thanks to Ana Pungartnik for the copyright research.

Printed by Elanders Ltd, Newcastle upon Tyne

Based on the classic CGP style created by Richard Parsons.

Contents

✓ Use the tick boxes to check off the topics you've completed.

Module 1 (Development of Practical Skills) is tested in context throughout this book, alongside Modules 2 to 4.

Exam Advice

Good exam technique can make a big difference to your marks, so make sure you read this stuff carefully.

Get Familiar with the **Exam Structure**

If you're sitting the A-level in Biology rather than the AS, you'll be sitting a different set of exams to the ones described here. The exams will include the same types of questions though.

For **AS Biology**, you'll be sitting **two papers**.

Breadth in biology (Modules 1 to 4) **1 hour 30 minutes** **70** marks **50%** of your AS	A section of **multiple choice** questions, plus **short answer** questions.
Depth in biology (Modules 1 to 4) **1 hour 30 minutes** **70** marks **50%** of your AS	**Short answer** and **extended response** questions

1) As you can see, **both papers** test you on **Modules 1 to 4**. Module 1 relates to **practical skills**. The Module 1 theory is tested in context throughout this book, alongside Modules 2 to 4.
2) Your **maths skills** will also be tested in both papers.

The **Quality** of Your **Extended Response** will be **Assessed**

1) The **Depth in biology** paper will contain **one or more extended response questions**.
2) These questions are worth **6 marks** and require a **long answer**. They're shown with an **asterisk** (*) next to their number.
3) You'll be awarded marks for the **quality** of your extended response as well as the **content** of your answer, so your answer needs to:

Make sure your answer is legible too (the same goes for all your written answers).

- Be written in **continuous prose** (so use paragraphs, not bullet points).
- Have a **clear** and **logical structure**.
- Use **appropriate scientific language**.
- Show **good reasoning** — i.e. show that you have thought about and understood the question, and can justify your answer.
- Include information that's **relevant** to the question.

Manage Your Time Sensibly

1) The **number of marks** tells you roughly **how long** to spend on a question — you've got **just over a minute per mark** in each paper. So if you get stuck on a short question, it's sometimes worth **moving on** to another one, and then **coming back** to it if you've got time **at the end**.
2) Bear in mind, you might want to spend a **bit longer** per mark on the **extended response** questions, in which case you'll have to spend a bit less time on the short answer questions.

Command Words Tell You **What You Need** to do in a Question

Here are some of the **most common ones**:

Command word	What to do
Give / Name / State	Give a brief **one or two word** answer, or a short sentence.
Describe	Write about what something's like in detail. E.g. describe the structure of a cell.
Explain	**Give reasons** or **causes** for something.
Suggest	Use your **scientific knowledge** to work out what the answer **might** be.
Calculate	Work out the **solution** to a mathematical problem.

Some questions will ask you to answer '**using the information/data provided**' — if so, you must **refer** to the information, data or figure you've been given or you won't get the marks.

Cell Structure — 1

No, no, not that type of cell — though revision can feel like a prison. This section's quite interesting, really...

For each of the questions 1-3, give your answer by writing the correct letter in the box.

1 The micrograph below shows part of a cell.

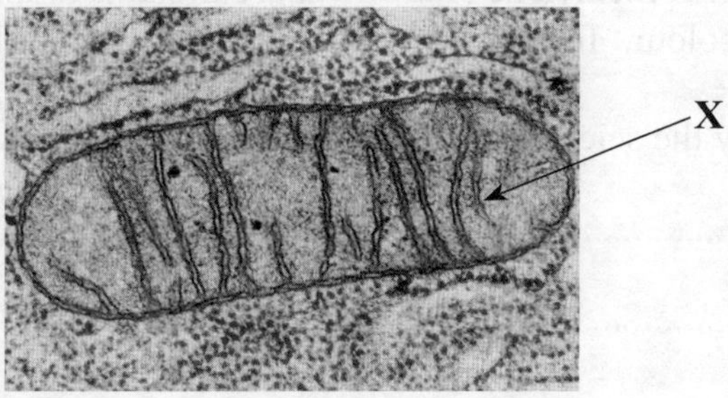

Which option, **A** to **D**, correctly identifies the structure labelled **X** in the micrograph?

A stroma **B** crista

C lamella **D** granum

Your answer [B]

(1 mark)

2 An organelle has the properties listed below.

- There are pairs of microtubules inside the structure.
- It is found on the surface of a cell.
- It is surrounded by a membrane.

Which organelle, **A** to **D**, is being described?

A cell wall **B** chloroplast

C flagellum **D** centriole

Your answer [C]

(1 mark)

3 The table below shows the role of three organelles in the production and secretion of proteins from a cell.

Which of the rows in the table, **A** to **D**, correctly identifies the functions of the organelles?

	Ribosome	Rough endoplasmic reticulum	Golgi apparatus
A	Packages proteins for transport.	Folds and processes proteins.	Makes proteins.
B	Makes proteins.	Packages proteins for transport.	Folds and processes proteins.
C	Folds and processes proteins.	Packages proteins for transport.	Makes proteins.
D	Makes proteins.	Folds and processes proteins.	Packages proteins for transport.

Your answer [B]

(1 mark)

4 A scientist observed the blood cells in a sample of blood. It is possible to distinguish between white and red blood cells because white blood cells contain a nucleus and red blood cells do not.

(a) Describe the function of a cell's nucleus.

....

....

(1 mark)

The scientist added eosin Y, methylene blue and azure B stains to the sample. Eosin Y stains the cytoplasm of cells an orange/pink colour. Together, methylene blue and azure B stain cell nuclei a blue/purple colour.

(b) Suggest and explain why the scientist used these stains to observe white blood cells in the sample.

....

....

....

(2 marks)

The scientist viewed the blood sample using a light microscope.

(c) Describe how the scientist could have used a light microscope to view cells on a prepared slide containing the blood sample.

Start on the lowest objective lens then use the course adjustment to move the stage so the sample is roughly focused then use the fine focus until there is a clear image produced

....

....

(4 marks)

(d) (i) **Figure 1** shows an image of a white blood cell in the blood sample that the scientist observed. She measured the size of the cell to be 12 µm, using an eyepiece graticule. The scientist used the line (**X** to **Y**) to measure the size of the cell. Calculate the magnification of the image shown in **Figure 1**. Show your working.

If you need to measure something in an exam, do it in millimetres. This'll make it easier to convert to other units. For example, micrometres = mm × 1000 and nanometres = µm × 1000.

Figure 1

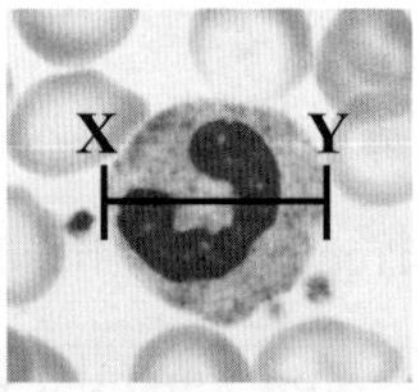

12 ÷ 1000 = 0.012
15 ÷ 0.012 = 1250

Magnification = ×1250

(2 marks)

(ii) The scientist calculated a percentage error of 0.4% when measuring cell size with the eyepiece graticule. Use this information to calculate the uncertainty of this measurement. Give your answer to one significant figure.

Uncertainty = ... µm

(2 marks)

5 Sperm cells are specialised for their function of delivering genetic material to the egg.

Figure 2 shows the structure of a sperm cell from an animal.

Figure 2

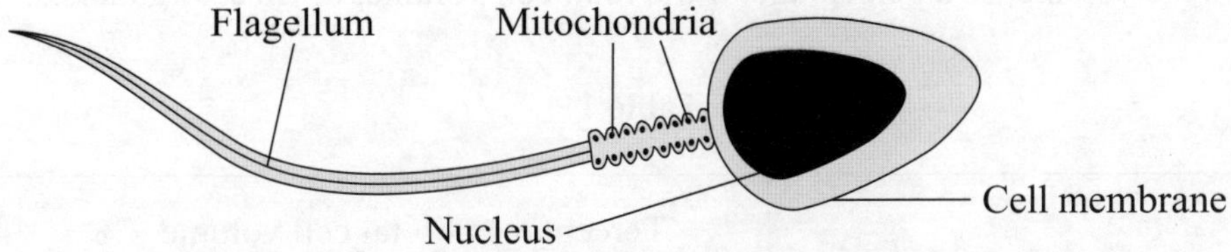

(a) Using **Figure 2**, give **one** similarity and **one** difference between a sperm cell and a bacterium.

....

....

....

....

(2 marks)

(b) (i) Suggest why the mitochondria are located close to the sperm cell's flagellum.

....

....

(1 mark)

(ii) A scientist wanted to observe the mitochondria in a sample of sperm.

Which type of microscope should the scientist use to study the internal structures of mitochondria? Explain your answer.

....

....

....

(2 marks)

(c) Dynein proteins are responsible for causing microtubules to bend.
A mutation in the *DNAH1* gene causes dynein proteins to no longer be expressed.
Suggest and explain how this may lead to male infertility.

....

....

....

(2 marks)

EXAM TIP

For questions on cell structure, you can be given a micrograph. It can be tricky to interpret these and to spot the different cell structures, but they are there. Trust me. To prepare yourself for the exam, make sure you learn what all the different cell structures look like.

Score

21

Cell Structure — 2

1 A team of scientists studied the organelles in **two** types of cell (**A** and **B**) taken from the body tissues of a eukaryotic organism.

Table 1 shows the volume, as a percentage of the total cell volume, of **three** organelles.

Table 1

Organelle	Percentage of total cell volume / %	
	Cell type **A**	Cell type **B**
Lysosomes	4	1
Rough endoplasmic reticulum	8	16
Nucleus	7	7

(a) The relative volume of the nucleus is the same in both types of cell. Suggest why.

....

....

(1 mark)

(b) The role of one of the two cell types is to ingest invading pathogens, and the other is to secrete enzymes.

Use **Table 1** to determine which of these two roles is carried out by cell type **A** and which is carried out by cell type **B**. Explain your answers.

A's job is to ingest invading pathogens because it has more Lysosomes which are needed to break down the pathogen. Cell B's job is to secrete enzymes because it has, more RER than cell A. This is covered in ribosomes synthesis proteins

(4 marks)

(c) Two other organelles that can be found in eukaryotes are mitochondria and chloroplasts. Compare the structure and function of these organelles to give **two** differences.

1. ....

....

2. ....

....

(2 marks)

2 A student observed the cells from the skin of an onion under a light microscope. To do so, he prepared a wet mount slide. He then added a stain.

(a) (i) Explain how the student could have added a stain to his wet mount slide.

....

....

....

(2 marks)

(ii) Give **one** safety precaution the student should have taken when preparing the slide.

....

(1 mark)

(b) The student used an eyepiece graticule to calculate the size of some of the onion cells. **Figure 1** shows the student's eyepiece graticule and stage micrometer. The stage micrometer measures in millimetres.

Figure 1

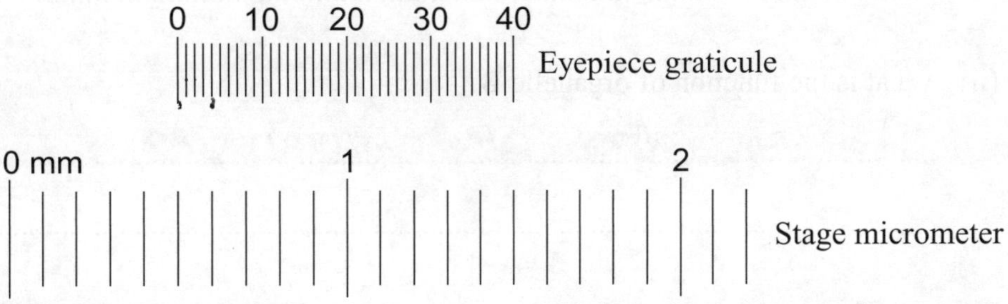

(i) Use **Figure 1** to calculate the size of **one** division on the student's eyepiece graticule, in micrometres. Show your working.

0.1÷4 = 0.025 mm

0.025 × 1000 = 25

Answer = 25 µm

(2 marks)

(ii) The student increased the magnification, so he needed to recalibrate the eyepiece graticule. Explain why the student needed to recalibrate the graticule.

....

(1 mark)

Another student calculated the size of an onion cell from an image. **Figure 2** shows the cell at × 100 magnification.

Figure 2

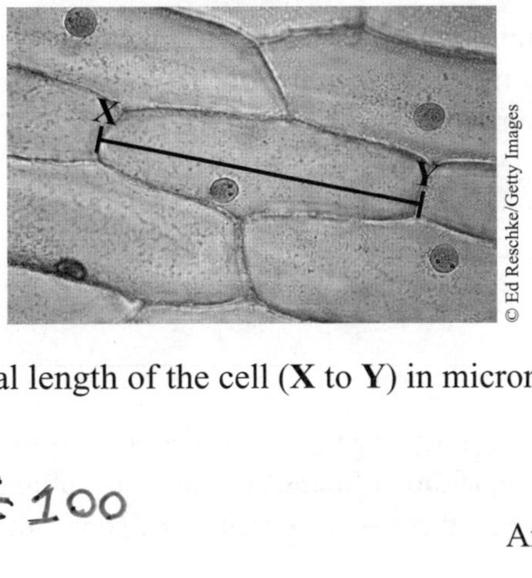

(c) Using **Figure 2**, calculate the real length of the cell (**X** to **Y**) in micrometres. Show your working.

36×1000 = 36000 ÷ 100

Answer = 360 µm

(2 marks)

3 Microtubules are important components of cell structure. **Figure 3** shows the microtubules present in the cross-sections of **two** different organelles (**A** and **B**).

Figure 3

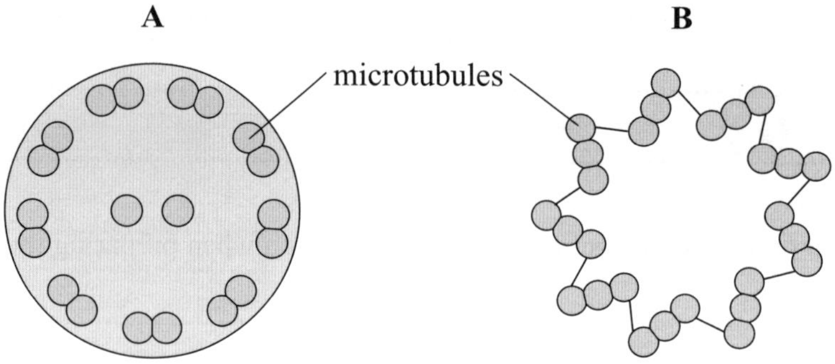

(a) (i) Suggest the identity of organelle **A**. Explain your answer.

Its a flagellum cause there are lots of microtubles

(2 marks)

(ii) What is the function of organelle **B**?

The seperation of chromosomes

(1 mark)

(b) The breakdown of the cytoskeleton in nerve cells has been linked to Parkinson's disease. People with Parkinson's disease can experience a lack of control of their muscles.

Nerve cells transmit nervous impulses to other cells, such as muscles, using neurotransmitter proteins. These proteins are released from vesicles transported to the surface of nerve cells.

(i) Suggest **two** reasons why cytoskeleton degradation in nerve cells could lead to a lack of muscle control.

1. ....

2. ....

(2 marks)

(ii) Abnormal function of certain organelles has been linked to some symptoms of Parkinson's disease. Suggest **one** organelle that may lead to some of these symptoms if it showed abnormal function in nerve cells. Explain your answer.

If mitochondria don't work cells dont get enough energy and so die.

(1 mark)

EXAM TIP

Maths always manages to worm its way in. You really need to make sure you learn some formulas, like the one for magnification (magnification = size of image ÷ size of real object). And make sure you *really* know them — you need to be confident rearranging them.

Score

21

Biological Molecules — 1

Biological molecules are essential for living organisms — answering these questions is essential for your revision.

1 Which of the rows in the table, **A** to **D**, correctly identifies a phosphate ion?

	Chemical symbol	Cation	Anion
A	PO_4^{3+}		✓
B	PO_3^{4-}	✓	
C	PO_4^{3-}	✓	
D	PO_4^{3-}		✓

Your answer D

(1 mark)

2 The diagram below shows the structure of a molecule, called mucin.

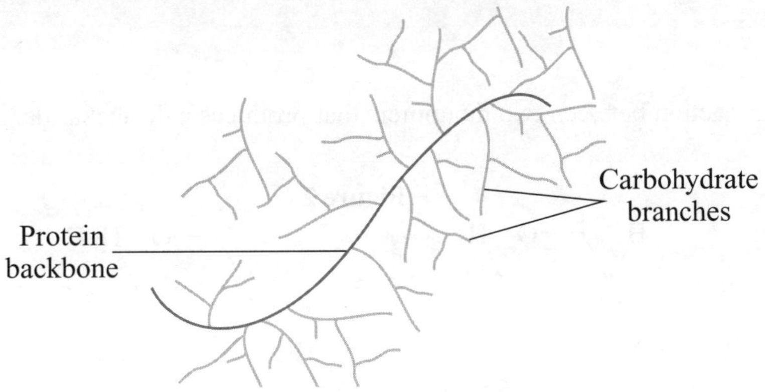

Which option, **A** to **D**, names the type of bonds present in a mucin molecule?

A glycosidic and ester bonds
B peptide and ester bonds
C glycosidic and peptide bonds
D peptide bonds only

Your answer C

(1 mark)

3 The table below shows information on four different molecules found in a cell.

Which of the rows in the table, **A** to **D**, describes the molecule that is the main component of cell membranes?

	Atoms in molecule	Variable group(s) present in molecule?
A	C, H, O, P	Yes
B	C, H, O	Yes
C	C, H, O	No
D	C, H, O, N	Yes

Your answer A

(1 mark)

4 **Figure 1** shows a polymer.

Figure 1

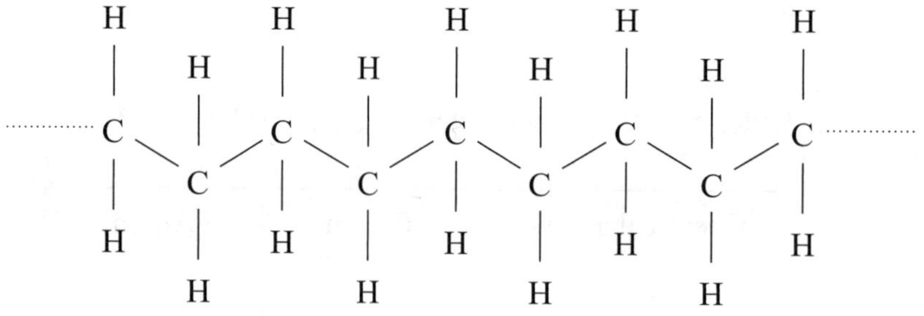

(a) What is a polymer?

....

(1 mark)

(b) (i) Draw a circle around a single monomer in **Figure 1**.

(1 mark)

(ii) Give **two** types of monomer found in biological molecules.

1. ....

2. ....

(1 mark)

5 **Figure 2** shows a reaction between two monomers that produces a disaccharide.

Figure 2

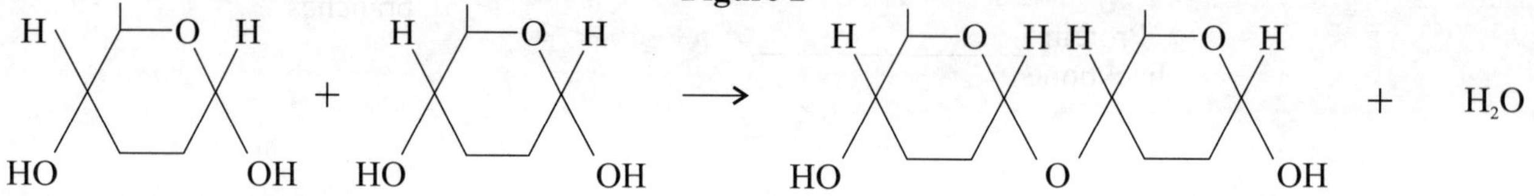

(a) (i) Name the monomers shown in **Figure 2**.

....

(1 mark)

(ii) Name the disaccharide produced in **Figure 2**.

....

(1 mark)

(iii) State the number of carbon atoms in the disaccharide produced in **Figure 2**.

Number of carbon atoms = ...

(1 mark)

(b) Disaccharides can be broken down.
Describe this reaction.

The two H's and the O are added back into the bond meaning that each bond has a single H and a OH

(3 marks)

6 Three molecules were tested and found to be proteins.

(a) Describe a test that can be carried out to determine whether or not a molecule is a protein. Include details of a positive result in your answer.

.....

.....

.....

(2 marks)

Further analysis of two of the proteins showed that they had slightly different properties.

Protein **A** was soluble and contained four iron-containing prosthetic groups.
Protein **B** was roughly spherical in shape and soluble.

(b) (i) Suggest the name of protein **A**.

.....

(1 mark)

(ii) It was concluded that protein **B** was insulin. Explain why solubility is important for the function of insulin.

.....

.....

(1 mark)

(iii) Explain how the structure of insulin makes it soluble.

.....

.....

.....

(2 marks)

(iv) The third protein was a fibrous protein, extracted from a sample of skin. Suggest **two** properties of the fibrous protein.

1. .....

2. .....

(2 marks)

7 Animals living in hot, dry climates have developed behaviours that help them keep cool.

Kangaroos have been observed licking saliva onto their forearms in hot weather.

(a) Using your knowledge of the properties of water, explain why this behaviour helps the kangaroos to keep cool.

The saliva contains water so the latent heat of vaporisation means that heat energy is released as the water evaporates

(2 marks)

Koalas have been observed to hug trees in hot weather.
This is thought to be because the trunks of trees are usually cooler than the surrounding air.

(b) (i) Tree trunks contain a lot of water.
Explain how this could contribute to the tree trunks being cooler than the surrounding air.

When the water evaporates the heat energy is released meaning that the tree trunk will cool down

(2 marks)

(ii) With reference to the structure and bonding of water molecules, explain how water is able to flow up a tree trunk, from the roots to the leaves.

Because of cohesion the water is pulled up.

(3 marks)

8 Proteins are polymers of amino acids. **Figure 3** shows the amino acid alanine.

Figure 3

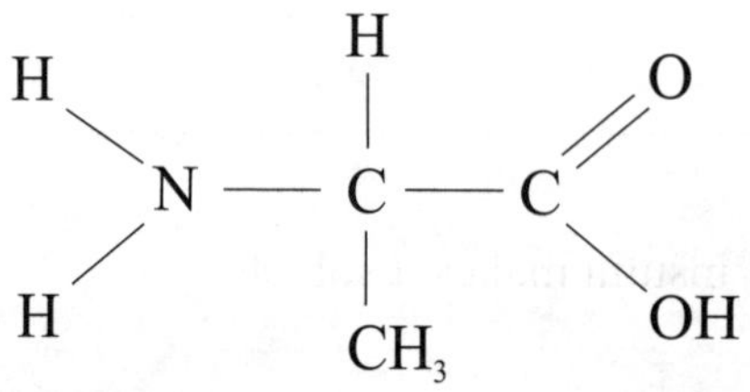

(a) (i) On **Figure 3**, circle and label the carboxyl group, the R group and the amino group.

(3 marks)

(ii) How is alanine different to the other 19 amino acids?

(1 mark)

(b) (i) Name the molecule released when a peptide bond is formed between **two** amino acids.

(1 mark)

(ii) Draw a diagram of the dipeptide formed from the reaction between **two** molecules of alanine. Label the peptide bond.

(2 marks)

EXAM TIP

You need to be really familiar with the structure of proteins, lipids and carbohydrates. You might get asked to identify a molecule or one of its groups from a diagram. Practice drawing these molecules out at home to help you visualise them — and don't be afraid to take a minute to quickly sketch out a molecule in your exam if it helps you to answer the question.

Score

34

Biological Molecules — 2

1 Proteins have four levels of structure.

Figure 1 shows part of the secondary structure of a protein.

Figure 1

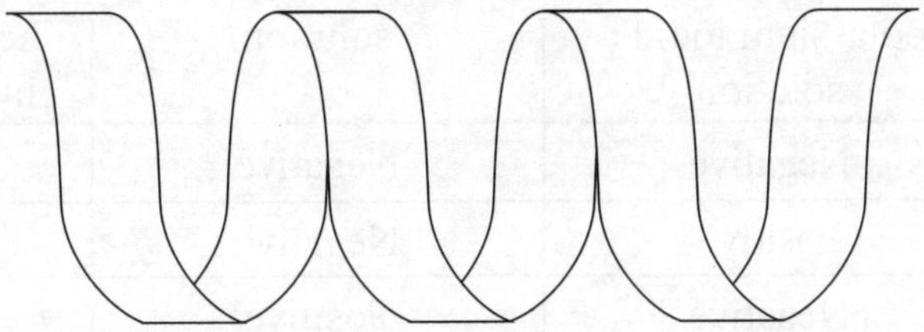

(a) (i) State which secondary structure is shown in **Figure 1**.

....

(1 mark)

(ii) Compare the bonding in the secondary and tertiary structures of a protein.

....

....

....

(2 marks)

Catalase is a globular, conjugated protein with a quaternary structure.

Catalase is an enzyme that catalyses the breakdown of hydrogen peroxide. In order to catalyse this reaction, the hydrogen peroxide molecules must bind to catalase's active site — this is an area of the protein with a specific 3D shape.

In humans, catalase is synthesised on ribosomes in the cytoplasm, but acts inside membrane-bound organelles known as peroxisomes.

(b)* Explain what the information above tells you about the structure of catalase and suggest how catalase's structure helps it to carry out its function.

....

....

....

....

....

....

....

....

....

(6 marks)

2 Three food samples (**A**, **B** and **C**), each containing carbohydrates, were tested using different techniques.

The results of these tests are shown in **Table 1**.

Table 1

Sample	Test Results		
	Test with iodine dissolved in potassium iodide solution	Test with Benedict's solution	Test with Benedict's solution (after heating with dilute hydrochloric acid)
A	Negative	Negative	Positive
B	Positive	Negative	Positive
C	Negative	Positive	Positive

(a) (i) Describe how to carry out a Benedict's test and what would indicate a positive result.

....

....

....

(2 marks)

The tests shown in **Table 1** allow the type of carbohydrate in each sample to be identified.

(ii) Using the information provided in **Table 1**, complete **Table 2** by placing a tick (✓) in the column that correctly identifies the type of carbohydrate present.

Table 2

Sample	Type of carbohydrate present		
	Reducing sugar	Non-reducing sugar	Starch
A			
B			
C			

(2 marks)

Two more samples were tested and found to contain the reducing sugar, glucose.

(b) (i) Describe how reagent test strips could be used to provide an indication of the glucose concentration in each sample.

....

....

....

(2 marks)

(ii) Give **one** other method of determining the glucose concentration in the two samples.

....

(1 mark)

3 **Figure 2** shows a type of biological molecule.

Figure 2

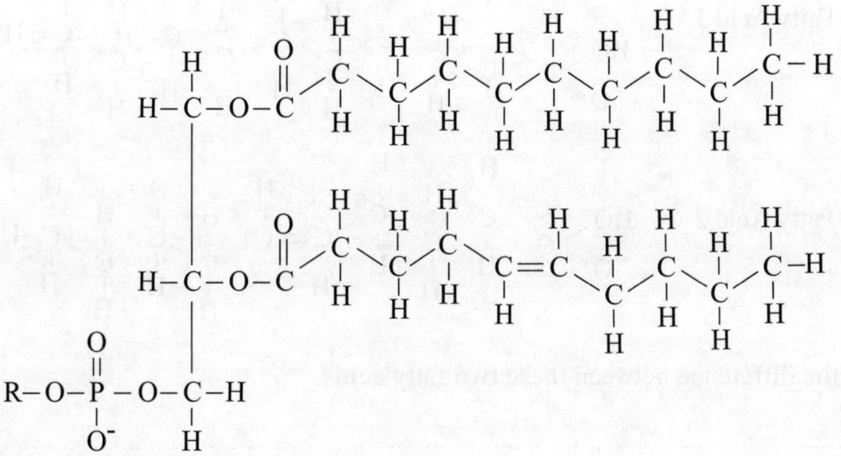

(a) Name the type of molecule shown in **Figure 2**.

....

(1 mark)

A droplet of these molecules was placed in water.
The molecules took the arrangement shown in **Figure 3**.

Figure 3

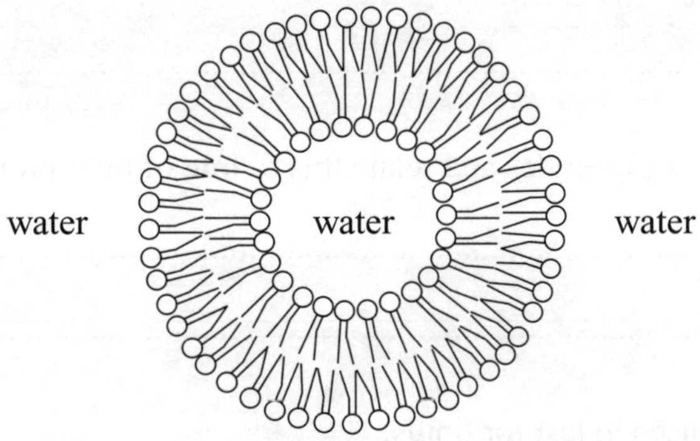

(b) (i) Explain why the molecules arranged themselves in this way.

....

....

....

....

(3 marks)

(ii) Describe **one** role that the molecules shown in **Figure 3** have in a cell.

....

....

....

(2 marks)

4 **Figure 4** shows two different fatty acids.

Figure 4

Fatty Acid 1

Fatty Acid 2

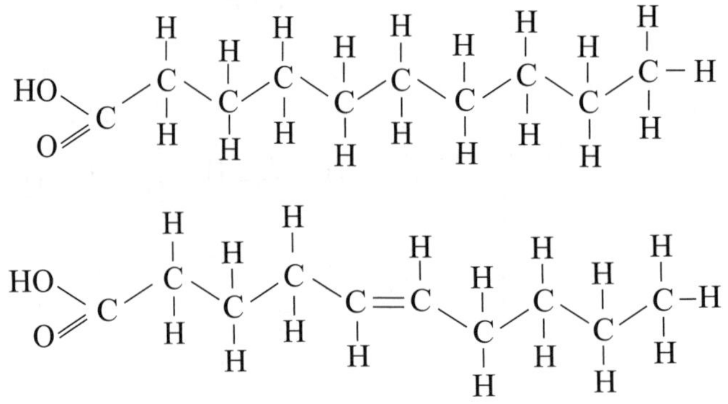

(a) (i) Explain the difference between these **two** fatty acids.

....

....

....

(2 marks)

Triglycerides contain fatty acids.

(ii) Describe how triglycerides are formed.

....

....

....

(3 marks)

(iii) Give **one** function of triglycerides and relate this to **one** of their properties.

....

....

(2 marks)

The emulsion test can be used to test for lipids.

(b) (i) Describe the emulsion test, including a positive result.

....

....

(2 marks)

(ii) An emulsion is droplets of one liquid suspended in another liquid.
Using this information, explain why lipids give a positive result in the emulsion test.

....

....

(1 mark)

EXAM TIP

Always make sure you read the instructions in an exam question carefully. If a question asks you to answer by placing ticks in a table and you put crosses in the table instead, you won't get the marks — even if your crosses are in the correct places. Harsh, but you have been warned.

Score

32

Biological Molecules — 3

1 Starch and cellulose are polymers of glucose. Starch is made of alpha-glucose molecules and cellulose is made of beta-glucose molecules.

(a) (i) Outline how starch is formed from alpha-glucose molecules.

....

....

....

(2 marks)

(ii) Draw the structure of beta-glucose below and explain how it is different from that of alpha-glucose.

....

....

(2 marks)

(iii) The beta-glucose molecules allow cellulose to form long, straight chains with multiple hydrogen bonds between the chains. Explain how this makes cellulose well-suited to its function.

....

....

(2 marks)

Starch is a mixture of two polysaccharides of alpha-glucose, amylose and amylopectin. The structures of amylose (**A**) and amylopectin (**B**) are shown in **Figure 1**.

Figure 1

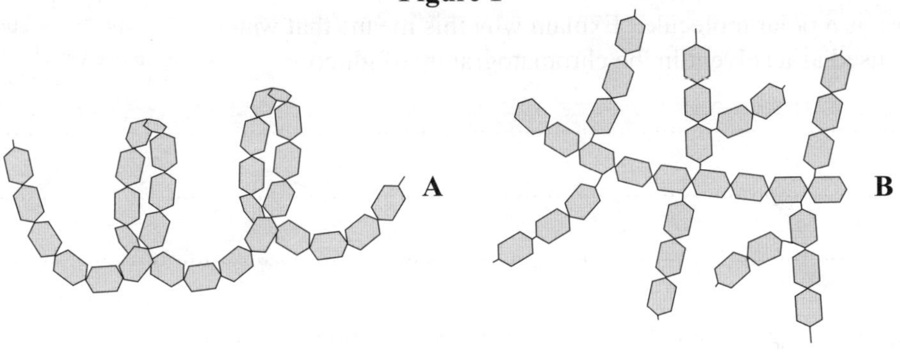

(b) Different starches are made up of different proportions of amylose and amylopectin. Using **Figure 1**, explain **one** advantage and **one** disadvantage of using amylose to store excess glucose, rather than amylopectin.

Advantage: ....

....

Disadvantage: ....

....

(2 marks)

2 A student was given three unknown samples of reducing sugar (**A**, **B**, **C**) to identify. She used paper chromatography to analyse the samples. The chromatogram produced is shown in **Figure 2**.

Figure 2

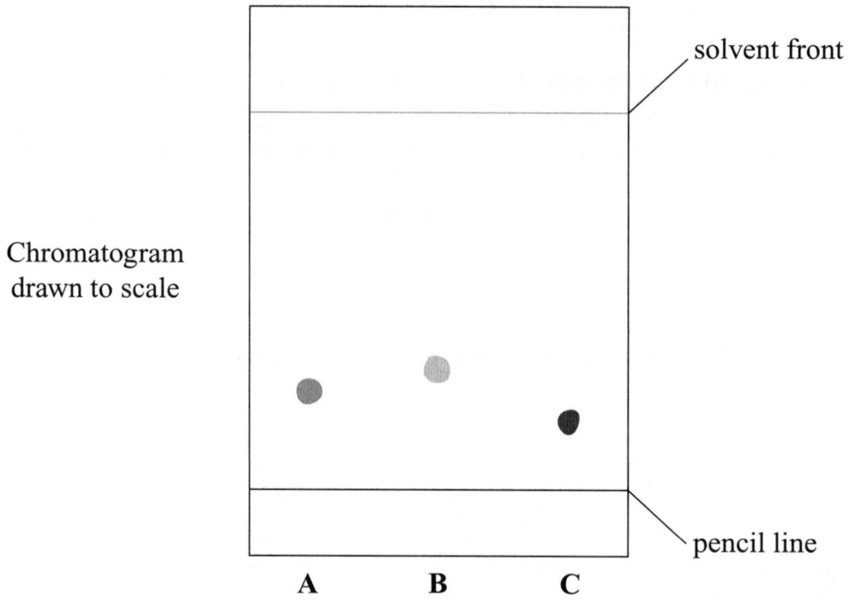

(a) (i) To produce her chromatogram, the student placed her prepared chromatography paper in a glass beaker containing a small amount of solvent. She then covered the beaker with a lid. Suggest why the student covered the beaker with a lid.

....

(1 mark)

(ii) Describe the role of a solvent in chromatography.

....

....

(1 mark)

(iii) Glucose is a polar molecule. Explain why this means that water can be used as a solvent in the chromatography of glucose.

Hint: think about the properties of water here.

....

....

....

(2 marks)

(b) (i) The R_f value of glucose in the solvent used by the student is 0.18.
Using **Figure 2**, determine which of the sugars (**A** to **C**) is glucose. Show your working.

Answer = ...

(2 marks)

(ii) Before the student could calculate the R_f values she dipped the chromatogram in Benedict's reagent. Suggest why the student needed to do this.

....

(1 mark)

3 A student wanted to investigate the concentration of glucose in different energy drinks. The student first produced four glucose solutions of different, known concentrations by diluting a 4.0 mM glucose solution. The volumes of distilled water and glucose solution he added to each test tube can be seen in **Table 1**. The total volume of each solution was 10 cm^3.

Table 1

Final concentration of glucose solution / mM	Volume of distilled water / cm^3	Volume of glucose solution / cm^3
4.0	0.0	10.0
	5.0	5.0
	7.5	2.5
	8.75	1.25

(a) Complete **Table 1** by filling in the three blank spaces with the correct final concentrations of glucose solution.

(1 mark)

(b) Describe how the student could use the glucose solutions he had prepared, along with a colorimeter, to determine the concentration of glucose in each energy drink.

....

....

....

....

....

(4 marks)

(c) (i) Give **one** negative control the student could use in his experiment.

....

(1 mark)

(ii) Explain why a control would be used in this experiment.

....

....

(1 mark)

(d) Some of the energy drinks investigated were coloured, while others were colourless. Explain why this may have reduced the accuracy of the student's results.

....

....

....

(2 marks)

EXAM TIP

Examiners love to test your maths skills, so there'll be several calculation questions in the exams. If a calculation question asks you to show your working, make sure you do. That way, even if you don't get the right answer, you might still pick up a mark for using the correct method.

Score

24

Nucleotides and Nucleic Acids — 1

Nucleotides are the building blocks of life — they join to form DNA and RNA, which ultimately control how every organism functions. This topic isn't too tricky, but here are some practice questions to keep you on your toes.

For each of questions 1-3, give your answer by writing the correct letter in the box.

1 Which of the rows in the table, **A** to **D**, correctly shows the molecules that make up ATP?

A	thymine	phosphate	adenine
B	phosphate	deoxyribose	adenine
C	ribose	phosphate	adenine
D	ribose	thymine	phosphate

Your answer ☐ **(1 mark)**

2 Which of the following options, **A-D**, is **not** needed to carry out a DNA precipitation reaction?

A detergent

B sodium chloride

C ethanol

D DNA helicase

Your answer ☐ **(1 mark)**

3 A DNA sequence before and after replication is shown below, on the left. The sequence is read from left to right. The table below shows which mRNA codons code for each type of amino acid.

DNA sequence before replication:
TACCTAGCT

DNA sequence after replication:
TACCTCGCT

Second mRNA base

First mRNA base	U	C	A	G	Third mRNA base
U	UUU - Phe	UCU - Ser	UAU - Tyr	UGU - Cys	U
	UUC - Phe	UCC - Ser	UAC - Tyr	UGC - Cys	C
	UUA - Leu	UCA - Ser	UAA - Stop	UGA - Stop	A
	UUG - Leu	UCG - Ser	UAG - Stop	UGG - Trp	G
C	CUU - Leu	CCU - Pro	CAU - His	CGU - Arg	U
	CUC - Leu	CCC - Pro	CAC - His	CGC - Arg	C
	CUA - Leu	CCA - Pro	CAA - Gln	CGA - Arg	A
	CUG - Leu	CCG - Pro	CAG - Gln	CGG - Arg	G
A	AUU - Ile	ACU - Thr	AAU - Asn	AGU - Ser	U
	AUC - Ile	ACC - Thr	AAC - Asn	AGC - Ser	C
	AUA - Ile	ACA - Thr	AAA - Lys	AGA - Arg	A
	AUG - Met	ACG - Thr	AAG - Lys	AGG - Arg	G
G	GUU - Val	GCU - Ala	GAU - Asp	GGU - Gly	U
	GUC - Val	GCC - Ala	GAC - Asp	GGC - Gly	C
	GUA - Val	GCA - Ala	GAA - Glu	GGA - Gly	A
	GUG - Val	GCG - Ala	GAG - Glu	GGG - Gly	G

A mutation occurred in the DNA sequence during replication. Which of the following, **A-D**, describes the result of the mutation when the corresponding mRNA sequence is translated?

A A serine (Ser) amino acid is replaced with an arginine (Arg) amino acid.

B There is no change to the amino acid sequence.

C An asparagine (Asn) amino acid is replaced with a glycine (Gly) amino acid.

D An aspartate (Asp) amino acid is replaced with a glutamate (Glu) amino acid.

Your answer ☐ **(1 mark)**

4 RNA molecules are single-stranded polynucleotides.

Figure 1 shows part of the structure of an RNA molecule.

Figure 1

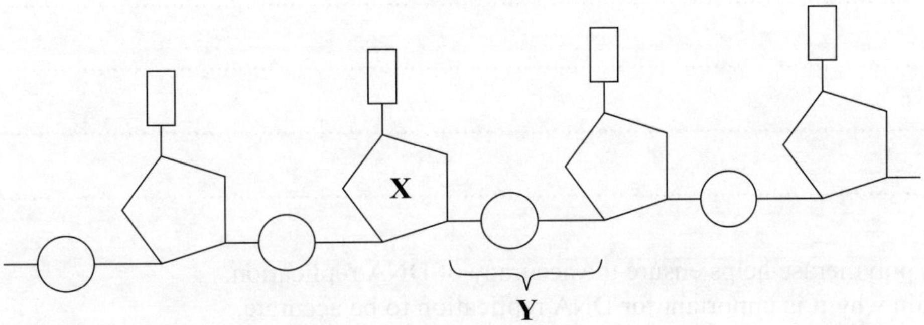

(a) (i) Name the structure labelled **X** in **Figure 1**.

....

(1 mark)

(ii) Name the bond labelled **Y** in **Figure 1**.

....

(1 mark)

Some RNA molecules are capable of folding into structures known as stem-loops. An example of a stem-loop structure is shown in **Figure 2**.

Figure 2

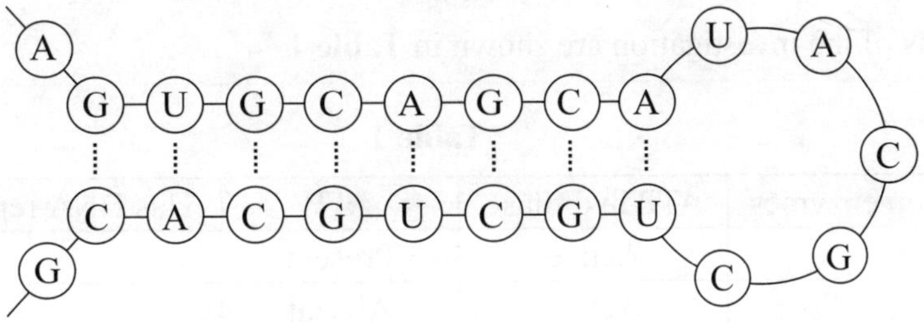

(b) (i) Looking at the sequence of the structure shown in **Figure 2**, explain how you can tell that this is part of an RNA molecule and not a DNA molecule.

....

....

(1 mark)

(ii) Using your knowledge of how DNA molecules can form a double helix, explain how the stem-loop structure shown in **Figure 2** is formed.

....

....

....

....

(3 marks)

(iii) Calculate the percentage of purine bases shown in **Figure 2**.

percentage of purine bases = ... %

(1 mark)

5 A scientist is investigating the role of enzymes in DNA replication.

(a) (i) Describe the roles that the enzymes DNA helicase and DNA polymerase play in DNA replication.

....

....

....

....

(3 marks)

(ii) DNA polymerase helps ensure the accuracy of DNA replication.
Explain why it is important for DNA replication to be accurate.

....

....

....

....

(3 marks)

The scientist mixes a bacterial DNA sample with the enzymes and substrates required for DNA replication. He does this in both the presence and the absence of ATP, and using active and inactive ATP hydrolase. ATP hydrolase is the enzyme that catalyses the breakdown of ATP. The scientist then measures the amount of DNA produced to determine whether DNA replication has taken place.

Some of the results of the investigation are shown in **Table 1**.

Table 1

DNA replication enzymes	ATP hydrolase	ATP	Has DNA replication occurred?
Present	Active	Present	Yes
Present	Active	Absent	No
Present	Inactive	Present	No
Present	Inactive	Absent	No

(b) (i) Describe the results shown in **Table 1**.

....

....

(1 mark)

(ii) Suggest an explanation for the results shown in **Table 1**.

....

....

....

(2 marks)

EXAM TIP

If you have to interpret the results of an unfamiliar experiment in the exam, don't panic — just make sure you read the method carefully. There will always be some important information there that will help you understand the results. If you can take a minute to piece it together with your existing biological knowledge, then you're sure to come out on top.

Score

19

Nucleotides and Nucleic Acids — 2

1 Leigh syndrome is a metabolic disorder that affects the central nervous system.
It can be caused by a mutation in the MT-ATP6 gene.

Figure 1 shows **one** of the mutations in the MT-ATP6 gene that can cause Leigh syndrome.
Table 1 contains some of the DNA codons that code for particular amino acids.

Figure 1

Original gene: CAACCAATAGCCC**T**GGCCGTA

Mutated gene: CAACCAATAGCCC**G**GGCCGTA

Table 1

Amino acid	DNA codon
Isoleucine	ATT, ATC, ATA
Glutamate	GAA, GAG
Leucine	CTG, TTA, TTG
Methionine	ATG
Valine	GTT, GTC, GTA, GTG
Arginine	CGG, AGA
Alanine	GCT, GCC, GCA, GCG

(a) Give **one** piece of evidence from **Table 1** that shows the genetic code is degenerate.

....

....

(1 mark)

(b) (i) Describe the effect that the mutation shown in **Figure 1** will have on the mRNA sequence produced from the MT-ATP6 gene.

Hint: Figure 1 will be easier to read and interpret if you mark off every three bases with a slash, e.g. CAA/CCA/...

....

....

(1 mark)

(ii) Using **Table 1** for reference, describe the effect that the mutation shown in **Figure 1** will have on the primary structure of the protein produced from the MT-ATP6 gene.

....

....

(1 mark)

(c) Complete the following passage to show how mRNA from the MT-ATP6 gene is translated into a protein.

The mRNA attaches to the ribosome. tRNA molecules carry .... to the ribosome. tRNA molecules attach to the mRNA via .... rRNA in the ribosome catalyses the formation of a .... between two .... The ribosome moves along the mRNA, producing a polypeptide chain. The process continues until a .... on the mRNA is reached.

(5 marks)

2 A team of scientists have developed a new drug. They want to investigate how the levels of three different mRNA molecules change when eukaryotic cells are treated with the drug.

To do so, they extract all of the mRNA from a sample of eukaryotic cells, and use an enzyme to convert the mRNA sequence into a complementary DNA (cDNA) sequence. They then determine the cDNA sequence and use it to identify the mRNA molecules that are present. The scientists analyse the data from treated and untreated cells to determine the level of mRNA expression.

(a) Outline how mRNA is produced from DNA by RNA polymerase.

....

....

....

....

....

....

(4 marks)

The results of the scientists' experiment are shown in **Figure 2**.

Figure 2

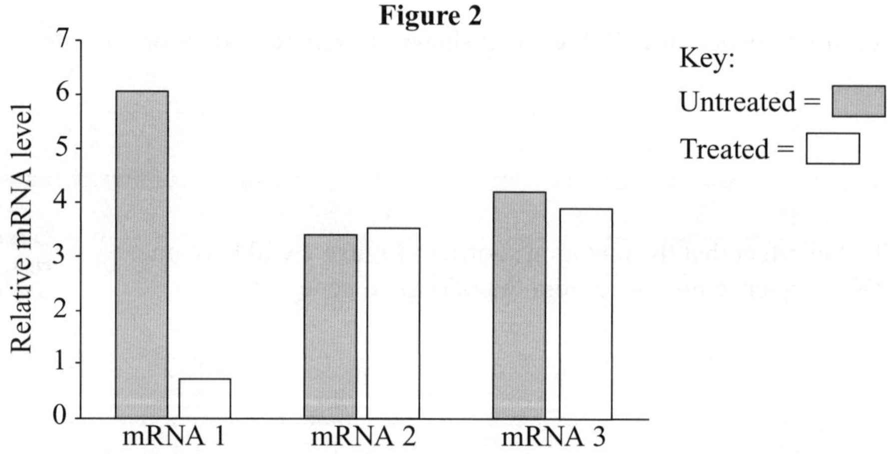

The scientists hypothesise that the new drug has two possible methods of action:
Method 1: By preventing RNA polymerase from working.
Method 2: By destroying particular mRNA sequences.

(b) (i) With reference to **Figure 2**, explain why the drug cannot be acting via Method 1.

....

....

....

(2 marks)

(ii) Explain how the results shown in **Figure 2** can be explained if the drug acts via Method 2.

....

....

....

(2 marks)

3* The scientists Meselson and Stahl carried out an experiment which showed that DNA replication is a semi-conservative process. The experiment used two types of nitrogen — heavy nitrogen and light nitrogen.

Meselson and Stahl grew *E. coli* bacteria in a nutrient broth which contained heavy nitrogen. Each time the bacteria divided, they replicated their DNA and took up heavy nitrogen from the broth to make new DNA nucleotides.

After several rounds of division, the heavy nitrogen was incorporated into all of the *E. coli* DNA. Meselson and Stahl then transferred the *E. coli* (bacterial generation 0) from the heavy nitrogen nutrient broth, to a nutrient broth containing only light nitrogen. They were left to divide several times.

Meselson and Stahl took a DNA sample from bacterial generation 0, and from the generations produced after each round of division in the light nitrogen broth. They then analysed the DNA to determine what percentage of the DNA contained only heavy nitrogen, both heavy and light nitrogen, or only light nitrogen. Some of the results of the experiment (bacterial generations 0-3) are shown in **Table 2**.

Table 2

Bacterial generation	Percentage of DNA with heavy nitrogen only	Percentage of DNA with both heavy and light nitrogen	Percentage of DNA with light nitrogen only
0	100	0	0
1	0	100	0
2	0	50	50
3	0	25	75

Using the information provided, explain how the experiment provides evidence that DNA replication is a semi-conservative process. You should include details of semi-conservative replication in your answer.

....

....

....

....

....

....

....

....

....

....

....

....

....

(6 marks)

EXAM TIP

Question 3* is a tricky one, there's no doubt about it. Not only does it ask you to interpret some unfamiliar experimental results, it also awards marks for the quality of your extended response. This means your answer should be well-structured, relevant and detailed. Take your time reading the information and think carefully about what your answer should include before you start.

Score

22

Enzymes — 1

Ahh, enzymes, now here's an interesting topic — these are important proteins that speed up lots of reactions. Most enzymes are really specific, and work best under optimal conditions. Now, let's see what you know...

For questions 1 and 2, give your answer by writing the correct letter in the box.

1 The rate of an enzyme-controlled reaction at **15 °C** is 0.8 $cm^3\ min^{-1}$.
The optimum temperature for the reaction is 40 °C. The reaction has a Q_{10} of 2.

Which of the following options, **A** to **D**, gives the rate of reaction at **25 °C**?

A 0.4 $cm^3\ min^{-1}$

B 0.8 $cm^3\ min^{-1}$

C 1.6 $cm^3\ min^{-1}$

D 2.4 $cm^3\ min^{-1}$

Your answer ☐

(1 mark)

2 The rate of a chemical reaction involving an enzyme was recorded over 60 seconds, as substrate concentration was increased. The results are shown in **Figure 1**.

Which of the figures, **A** to **D**, shows what the reaction curve would look like for 60 seconds of the same reaction, in the presence of a competitive inhibitor?

Figure 1

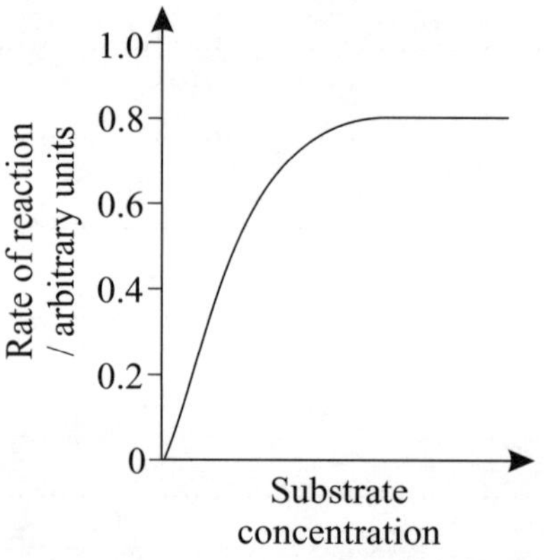

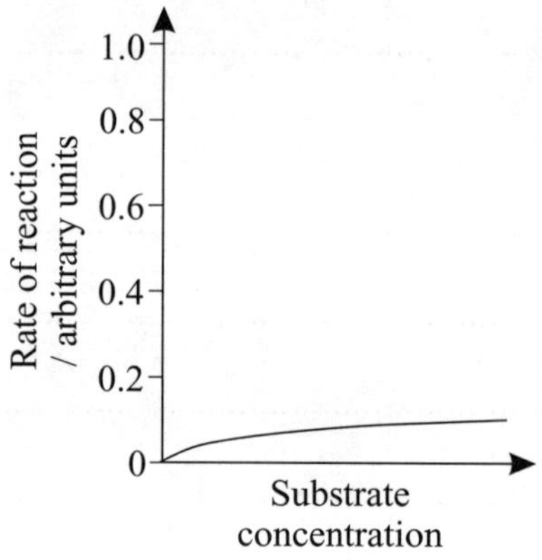

A

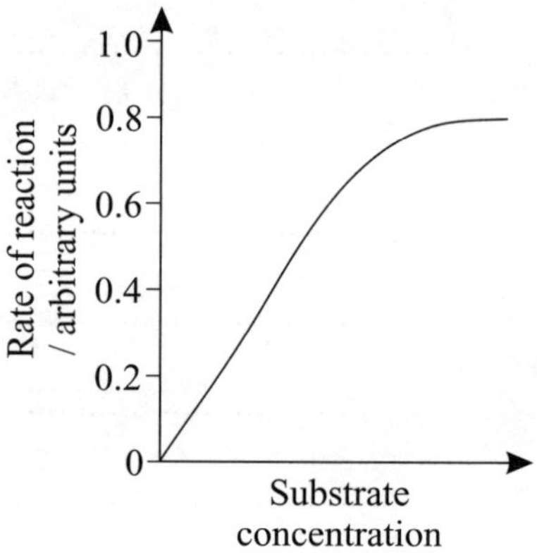

B

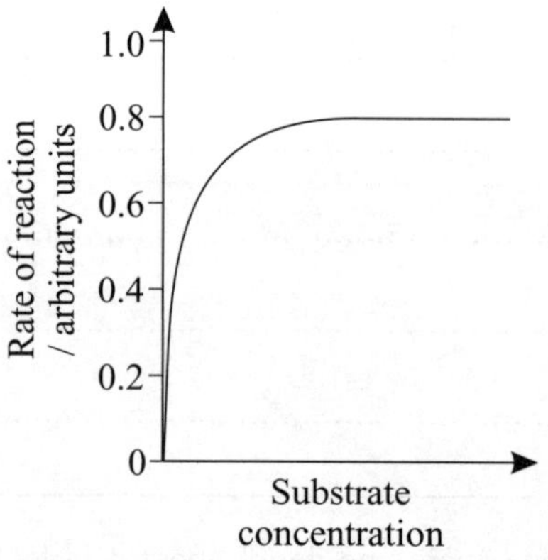

C

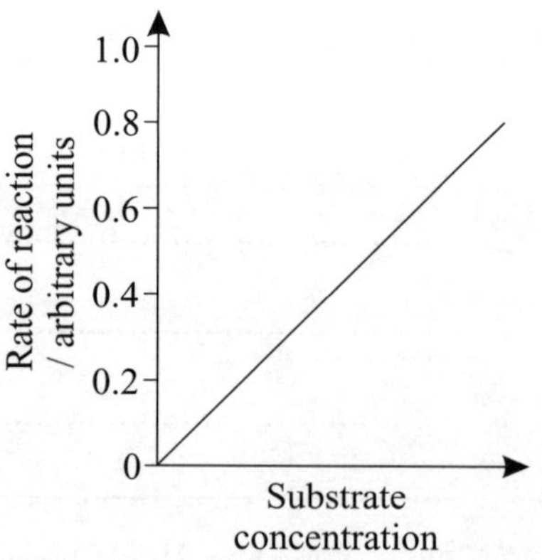

D

Your answer ☐

(1 mark)

3 **Figure 2** shows the activity of two different enzymes (**A** and **B**).
The enzymes are involved in respiration. One enzyme is from an insect that lives in the UK and one is from an insect that lives in a tropical climate.

Figure 2

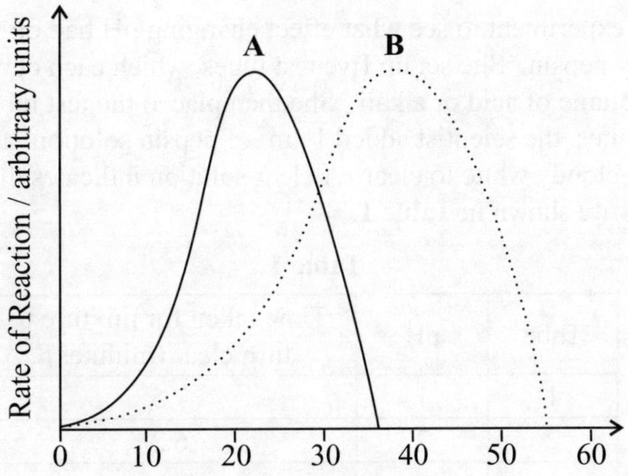

(a) Explain the shape of the curve for enzyme **A**.

....

....

....

....

(3 marks)

(b) Suggest which enzyme (**A** or **B**) is from the tropical insect. Explain your answer.

....

....

....

(2 marks)

Insecticides are chemicals that kill insects.
Scientists have developed an insecticide that works as a competitive inhibitor of enzyme **A**.

(c) Explain how the insecticide works.

> You need to use information from the introduction to question 3, as well as your knowledge of how competitive inhibitors work, to answer this question part.

....

....

....

....

....

(4 marks)

EXAM TIP

The main factors that influence the rate of an enzyme-controlled reaction are temperature, pH, enzyme concentration and substrate concentration. It's important you understand how each factor affects the rate, so that you can interpret graphs like the ones in questions 2 and 3.
You also need to learn how different types of inhibitor can affect enzyme-controlled reaction rates.

Score

11

Enzymes — 2

1 Pepsin is an enzyme that is found in the stomach and catalyses the breakdown of protein.

A scientist carried out an experiment to see what effect changing pH had on the breakdown of egg white, which contains protein, by pepsin. She set up five test tubes, which each contained 5 cm^3 of egg white suspension and a small volume of acid or alkali. She then placed the test tubes into a water bath set at 40 °C. After a few minutes, the scientist added 1 cm^3 of pepsin solution, and timed how long it took for the solutions to turn from cloudy white to clear. A clear solution indicates that all the protein has been broken down. The results are shown in **Table 1**.

Table 1

Tube	pH	Time taken for mixture to turn clear (minutes)
1	1	7.5
2	2	6.2
3	5	11.5
4	8	>15
5	11	>15

(a) Suggest why the temperature of the water bath was set to 40 °C.

....

....

(1 mark)

(b) The pH of the stomach ranges from pH 1.5 to pH 3.5.
The pH of the small intestine ranges from pH 6 to pH 7.
Use the results in **Table 1** to explain why pepsin is active in the stomach, but not in the small intestine.

....

....

....

....

....

(4 marks)

2 A student investigated how an enzyme-controlled reaction is affected by changes in enzyme concentration. The student used the enzyme catalase to break hydrogen peroxide down into water and oxygen.

(a) (i) The student used an extracellular source of catalase. In the human body, catalase is an intracellular enzyme. Explain the difference between extracellular enzymes and intracellular enzymes.

....

....

(1 mark)

(ii) Give **one** example of an enzyme that acts extracellularly in the human body.

....

(1 mark)

The results of the student's first experiment are shown in **Figure 1**.

Figure 1

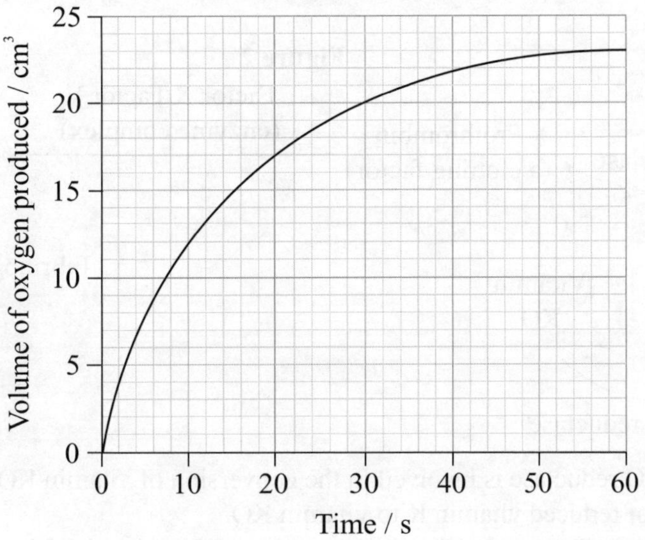

(b) Identify the dependent and independent variables in this investigation.

....

....

(1 mark)

The student collected the oxygen in a measuring cylinder submerged in water.
The measuring cylinder measured to the nearest 1 cm^3.

(c) Give the uncertainty of measurements associated with this measuring cylinder.

uncertainty = ... cm^3

(1 mark)

(d) Suggest **one** way that the student could have obtained more accurate results.

....

(1 mark)

(e) (i) Calculate the average rate of the reaction shown in **Figure 1** for the first 20 seconds.
Show your working and include appropriate units in your answer.

rate = ...

(2 marks)

(ii) Calculate the initial rate of the reaction shown in **Figure 1**. Show your working.

To calculate the initial rate, you need to draw a tangent.

rate = ...

(2 marks)

(f) Sketch on the same axes the curve you would expect if the experiment were carried out with a higher enzyme concentration.

(1 mark)

3 There are many enzymes and coenzymes involved in the mechanism of blood clotting. These act in a sequence to form a blood clot. Part of this sequence is shown in **Figure 2**.

Figure 2

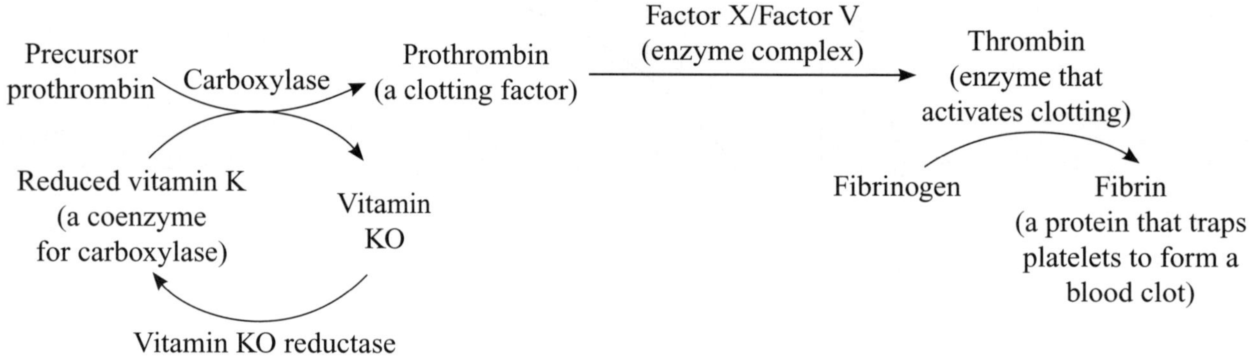

(a) The enzyme Vitamin KO reductase is involved in the conversion of vitamin KO to reduced vitamin K, but not the conversion of reduced vitamin K to vitamin KO.
Use the induced fit model of enzyme action to suggest an explanation for this.

....

....

....

....

(3 marks)

(b) Reduced vitamin K acts as a coenzyme for carboxylase. What is meant by the term 'coenzyme'?

....

....

(2 marks)

Blood clots form to help repair wounds.
However if a clot forms in an artery or vein, it can cause problems by reducing blood flow.

(c) Warfarin is a medication that can be used to reduce the risk of blood clots forming in a blood vessel. Warfarin interferes with the normal blood clotting mechanism by irreversibly inhibiting the action of the enzyme vitamin KO reductase.
With reference to **Figure 2**, explain why the action of warfarin could prevent blood clotting.

....

....

....

....

....

....

(4 marks)

EXAM TIP

Pen, pencil, ruler, calculator... Oh, that's just my shopping list for my exam. Sometimes examiners will ask you to draw, measure or calculate, so you need to make sure you've got all this stationery — a ruler that can measure in millimetres, a calculator, a pencil (for drawing graphs or diagrams) and a pen with black ink. In fact, make sure you've got a couple of spare pens too.

Score

24

Biological Membranes — 1

Cell membranes may look simple through a microscope, but nothing's as simple as it first seems. In fact, they're actually pretty complex and really important for cells. These questions will make sure your knowledge is tip top.

For each of questions 1-3, give your answer by writing the correct letter in the box.

1 The following options describe different types of movement across a cell membrane.

Which of the options, **A** to **D**, does **not** require the use of a transport protein or ATP?

A Enzymes being released from a cell via exocytosis.

B Glucose moving against its concentration gradient, from the intestinal lumen, into the cytoplasm of the cells of the intestines.

C Oxygen moving into the cell.

D Sodium ions diffusing into a cell from the extracellular fluid.

Your answer ☐

(1 mark)

2 A number of cells from a plant leaf are placed in a solution. The solution is hypotonic to the fluid inside the leaf cells.

Which of the following statements, **A** to **D**, describes what will happen to the leaf cells when they are placed in the solution?

A The cells will become flaccid.

B The cells will become turgid.

C The cells will burst.

D There will be no change in the cells.

Your answer ☐

(1 mark)

3 A student is investigating the effect of different factors on the rate of diffusion. He makes up some agar jelly containing red cabbage juice, and cuts it into cubes of 1 cm^3 and 3 cm^3. The student then places one of each cube size into separate beakers containing different concentrations of dilute ammonia solution. He then times how long it takes for the ammonia to diffuse into the agar cubes and turn them from red to green.

Which of the treatments, **A** to **D**, will result in the colour changing most rapidly?

A 1 cm^3 cube, ammonia concentration = 0.5 mol dm^{-3}

B 3 cm^3 cube, ammonia concentration = 0.5 mol dm^{-3}

C 3 cm^3 cube, ammonia concentration = 1.0 mol dm^{-3}

D 1 cm^3 cube, ammonia concentration = 1.0 mol dm^{-3}

Your answer ☐

(1 mark)

4 **Figure 1** models the arrangement of molecules in a typical cell membrane, observed from above.

Figure 1

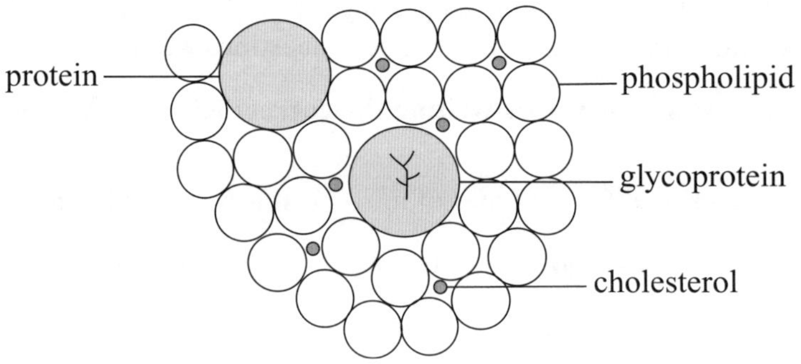

(a) (i) Describe the model illustrated in **Figure 1**.

....

....

(2 marks)

(ii) Explain the effect that a higher percentage of cholesterol would have on the model in **Figure 1**.

....

....

(2 marks)

(iii) State **two** roles of glycoproteins in the cell membrane.

1. ....

2. ....

(2 marks)

(iv) Describe what would happen to the structure of a cell membrane, if it was exposed to temperatures below 0 °C.

....

....

....

(2 marks)

(b) Cell membranes vary in structure due to the adaptation of specialised cells to their functions. For example, to conduct nerve impulses, a neurone cell relies upon the rapid movement of cations across its cell membrane.

(i) Suggest an adaptation that you might expect to observe in the cell membrane of a neurone cell. Give a reason for your choice.

....

....

(2 marks)

(ii) Low temperatures can reduce the speed at which nerve impulses are conducted. Suggest why.

....

....

....

(2 marks)

5 **Figure 2** shows part of the phospholipid bilayer in a cell-surface membrane.

Figure 2

Cell exterior — water potential = –0.5 kPa

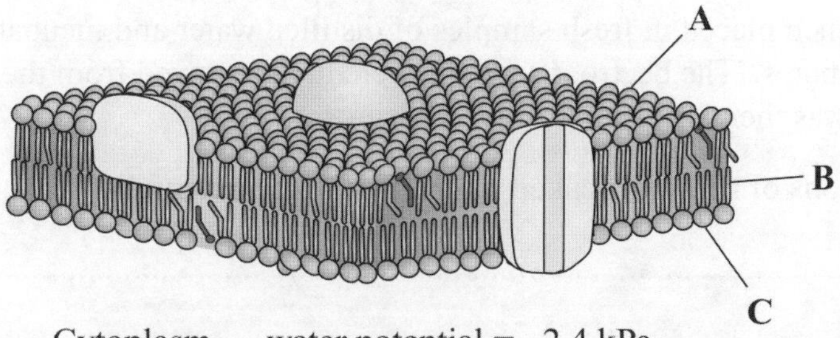

Cytoplasm — water potential = –2.4 kPa

(a) Which letter (**A-C**) represents the hydrophobic part of the phospholipid bilayer?

....

(1 mark)

(b) Explain how the phospholipid bilayer helps the membrane to carry out its function.

....

....

....

(2 marks)

(c) Using the information in **Figure 2**, describe the direction of movement of water across the cell membrane. Give a reason for your answer.

....

....

....

(2 marks)

6 Glucagon is a hormone that is released when there is not enough glucose in the blood. It binds to receptors on liver cells, causing the liver cells to break down stores of glycogen to glucose.

(a) Skeletal muscle cells also store glycogen.
Explain why glucagon affects liver cells, but not skeletal muscle cells.

....

....

....

(2 marks)

(b) REMD-477 is a drug that has a similar shape to glucagon. In the future, it might be used to treat diabetes because it is able to help keep the blood sugar level under control.
Suggest how REMD-477 helps keep the blood sugar level under control.

....

....

....

(2 marks)

7 Beetroot cells contain a vacuole. The vacuole contains red pigments called betalains, which are contained within the vacuole by a phospholipid membrane. A scientist wanted to investigate the effect of temperature on the permeability of this membrane.

Sections of beetroot were cut from the main plant and soaked in distilled water overnight. The cut sections were then placed in fresh samples of distilled water and incubated at different temperatures for 30 minutes. The beetroot sections were then removed from the water and discarded. Each sample of water was then analysed using a colorimeter.

(a) Why were the cut sections of beetroot soaked in distilled water overnight?

....

(1 mark)

(b) Suggest a negative control that could have been used in this investigation.

....

(1 mark)

Table 1 illustrates the results that were obtained from the colorimetry analysis. The percentage absorbance illustrates the proportion of transmitted light at blue/green wavelengths that was absorbed by the pigments in the water.

Table 1

Temperature / °C	Absorbance
20	0.05
30	0.16
40	0.35
50	0.60
60	0.73

(c) Explain, with reference to the structure of cell membranes, the results shown in **Table 1**.

....

....

....

....

....

(4 marks)

(d) A second investigation found that membrane permeability increased as the pH was decreased. Suggest an explanation for this.

....

....

....

(3 marks)

EXAM TIP

Examiners love a good practical question, so you'll need to brush up on terms like 'negative control'. When it comes to explaining results, be clear about what any table or graph is showing you — if the results for a colorimetry experiment show transmission of light rather than absorbance, the numbers will be the opposite way round to the ones above.

Score

33

Biological Membranes — 2

1 Mast cells are a type of immune system cell. They release a signalling molecule, called histamine, which triggers part of the immune response. Histamine is stored inside membrane-bound 'granules' in a mast cell's cytoplasm.

(a) Histamine is unable to diffuse across a mast cell's cell surface membrane. Suggest **one** reason why.

....

(1 mark)

(b) Suggest how mast cells release histamine into the blood.

....

....

....

(2 marks)

2 Plants contain a mixture of solutes. Depending on the relative concentrations of solutes inside a plant cell and its environment, water will move into or out of the cell by osmosis.

Some students wanted to investigate the water potential of white potato cells. To do so they incubated samples of white potato in different concentrations of sucrose solution. The mass of each sample was measured before and after the incubation. The change in mass was then calculated.

Figure 1 shows a calibration curve of the results.

Figure 1

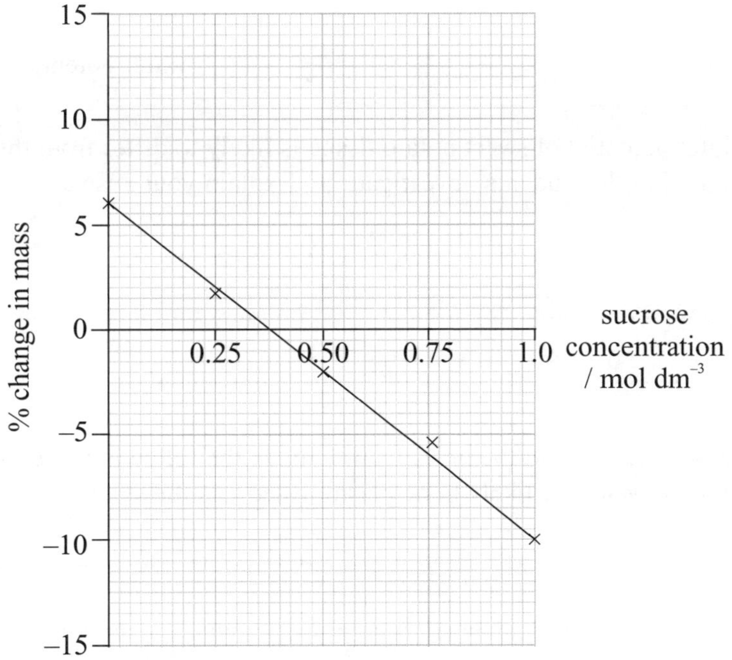

The students prepared the different concentrations of sucrose solution for their investigation using a stock solution of 1 mol dm^{-3} sucrose solution and distilled water.

(a) Complete **Table 1** to show the volumes of stock solution and water used to make up each concentration.

Table 1

Concentration of sucrose solution to be made up / mol dm^{-3}	Volume of 1 mol dm^{-3} sucrose solution used / cm^3	Volume of water used / cm^3	Final volume of solution to be made up / cm^3
1	20	0	20
0.75	15		20
0.5			20
0.25			20
0			20

(2 marks)

(b) Give **two** control variables for this investigation.

1. ....

2. ....

(2 marks)

Table 2 shows the relationship between sucrose concentration and water potential.

Table 2

Sucrose concentration / mol dm^{-3}	0.1	0.2	0.3	0.4	0.5	0.6	0.7	0.8	0.9	1.0
Water potential / kPa	–270	–540	–850	–1130	–1460	–1810	–2190	–2590	–3030	–3530

(c) Use **Table 2** and **Figure 1** to estimate the water potential of the potato tissue.
Show your working.

water potential = ... kPa

(2 marks)

(d) Suggest how the water potential of sweet potato tissue is likely to differ from the water potential of the white potato tissue used in the students' investigation. Explain your answer.

....

....

....

(2 marks)

(e) One of the students who carried out the investigation made the following statement:
'The potato cells in pure water are likely to be plasmolysed.' Is the student correct? Explain your answer.

....

....

(2 marks)

EXAM TIP

'Estimate' means you need to give an approximate value rather than calculate an exact answer. It needs to be a sensible estimate though, not just a stab in the dark guess, so you need to think about the best method to use to obtain your estimate.

Score

13

Cell Division and Cellular Organisation

Well, you're almost at the end of Module 2. Just one more section to get through. On you go then...

For each of questions 1-3, give your answer by writing the correct letter in the box.

1 Which of the following, **A** to **D**, correctly describes what is meant by the term 'homologous chromosomes'?

A Chromosomes in which the genetic information on each chromatid is identical.

B A pair of chromosomes in which each chromosome contains the same genes.

C Chromosomes that originate from the same parent cell prior to fertilisation.

D A pair of chromosomes in which each chromosome contains identical alleles.

Your answer ☐ **(1 mark)**

2 The human body has many different types of specialised cell, each of which is adapted to its function. Which of the rows, **A** to **D**, in the table below shows a correct adaptation for each cell type?

	Squamous epithelial cell	Neutrophil	Erythrocyte	Sperm cell
A	Joined to neighbouring cells via interlinking cell membranes	Has an acrosome containing digestive enzymes	Biconcave shape	Contains many lysosomes
B	Contains many lysosomes	Biconcave shape	Joined to neighbouring cells via interlinking cell membranes	Has an acrosome containing digestive enzymes
C	Joined to neighbouring cells via interlinking cell membranes	Contains many lysosomes	Biconcave shape	Has an acrosome containing digestive enzymes
D	Biconcave shape	Has an acrosome containing digestive enzymes	Joined to neighbouring cells via interlinking cell membranes	Contains many lysosomes

Your answer ☐ **(1 mark)**

3 The diagram below shows the main stages of the cell cycle. Which of the letters, **A** to **D** indicates the point in the cycle at which the attachment of chromosomes to the spindle is checked?

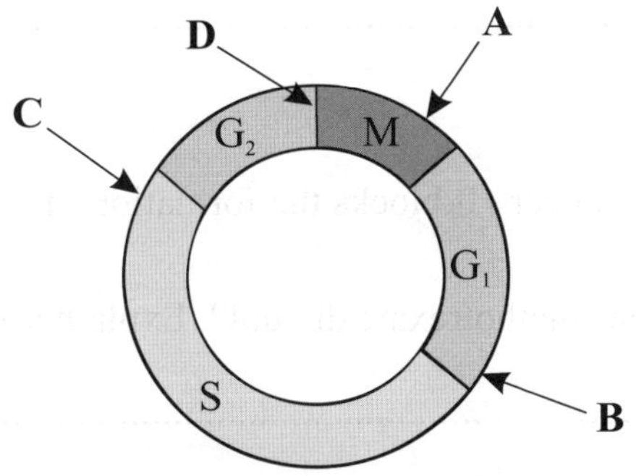

Your answer ☐ **(1 mark)**

4 A student investigated mitosis in an onion root tip.

(a) Explain why the student used the tip of the root for this investigation.

....

....

(1 mark)

(b) The student prepared a squash slide to view the cells.
Explain what is meant by a 'squash slide' and why the student prepared one for this investigation.

....

....

....

(2 marks)

The student's observations are shown in **Table 1**.

Table 1

Type of cell	Number of cells
Dividing	240
Non-dividing	80

(c) (i) Explain how the student was able to distinguish between dividing cells and non-dividing cells.

....

....

(1 mark)

(ii) Calculate the percentage of cells undergoing mitosis.

percentage of cells undergoing mitosis = ...%

(1 mark)

(d) The onion root tip contains stem cells.

(i) State **two** features of stem cells.

1. ....

2. ....

(2 marks)

(ii) Suggest **one** tissue type formed from the stem cells in a root tip.

....

(1 mark)

5 Methotrexate is a drug used to treat cancer. It blocks the formation of nucleotides within cells.

(a) (i) What stage of the cell cycle does methotrexate disrupt? Explain your answer.

....

....

(2 marks)

(ii) State why the cell cycle is unable to continue in cells that have been treated with methotrexate.

....

....

(1 mark)

(b) A low red blood cell count can be a side effect of taking methotrexate. Suggest why.

....

....

....

(2 marks)

6 A scientist is observing meiosis in the anthers of a diploid flowering plant.

(a) The scientist notes that some of the cells are undergoing anaphase II.

(i) Explain the difference between anaphase II and anaphase I.

....

....

....

(2 marks)

(ii) Outline the events that take place between the end of anaphase II and the completion of cell division.

....

....

....

(2 marks)

(b) Meiosis in the anthers produces pollen grains. A pollen grain is used to fertilise an ovum in the ovary of a female plant. Explain why meiosis is needed to produce a pollen grain.

....

....

....

(2 marks)

(c) Explain how meiosis creates genetic variation in plants.

....

....

....

....

....

....

(4 marks)

7 A scientist was studying the stages of the cell cycle.

The scientist used a microscope to observe some cells undergoing mitosis.
Figure 1 shows an image of one of these cells.

Figure 1

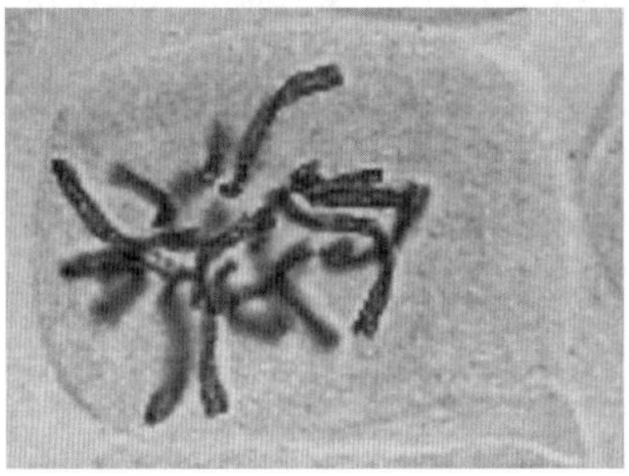

(a) Name the stage of mitosis shown in **Figure 1**. Explain your answer.

....

....

(2 marks)

Cyclins are proteins that play an important role in the cell cycle. A scientist recorded the concentration of two cyclins (**E** and **B**) during part of the cell cycle shown in **Figure 2**. He also recorded the mass of DNA present in the parent cell during this period (also shown in **Figure 2**).

Figure 2

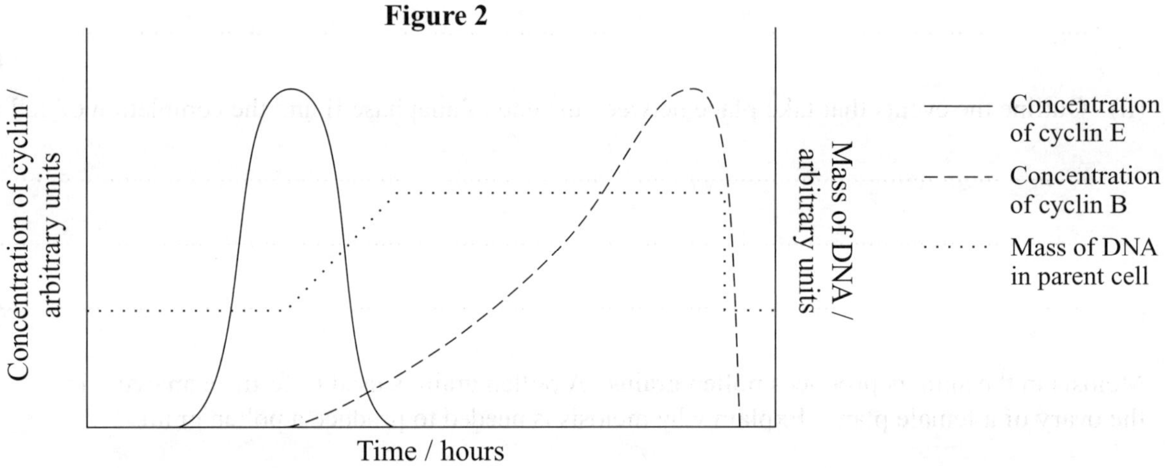

(b) Using the results shown in **Figure 2**, suggest the functions of cyclins **E** and **B** in the cell cycle.

....

....

....

....

....

....

(4 marks)

EXAM TIP

Exam questions on the cell cycle often describe a way that the normal cycle is altered and ask you to explain what effect this has. As long as you know what usually happens during the different stages of the cell cycle, you should be able to work out what happens if something changes. Make sure you're clued up on the different stages of interphase, mitosis and meiosis.

Score

32

Exchange and Transport — 1

If seeing 'Exchange and Transport Systems' got you excited about answering questions on the rail network, prepare to be disappointed. These are actually on how gases are exchanged in fish, insects, mammals and bacteria. Sorry.

For each of questions 1-3, give your answer by ticking the appropriate box.

1 The lungs contain various different cells and tissues, each of which has a different function. Which of the rows, **A** to **D**, in the table below shows the correct function for each component of the lungs?

	Goblet cell	Ciliated epithelium	Smooth muscle	Cartilage
A	Moves mucus away from the alveoli	Secretes mucus to trap microorganisms and dust	Provides support to the trachea and bronchi	Controls the diameter of the bronchioles
B	Secretes mucus to trap microorganisms and dust	Moves mucus away from the alveoli	Controls the diameter of the bronchioles	Provides support to the trachea and bronchi
C	Moves the mucus away from the alveoli	Provides support to the trachea and bronchi	Secretes mucus to trap microorganisms and dust	Controls the diameter of the bronchioles
D	Secretes mucus to trap microorganisms and dust	Controls the diameter of the bronchioles	Moves mucus away from the alveoli	Provides support to the trachea and bronchi

Your answer ☐ **(1 mark)**

2 The graph below shows the results of a spirometry test.

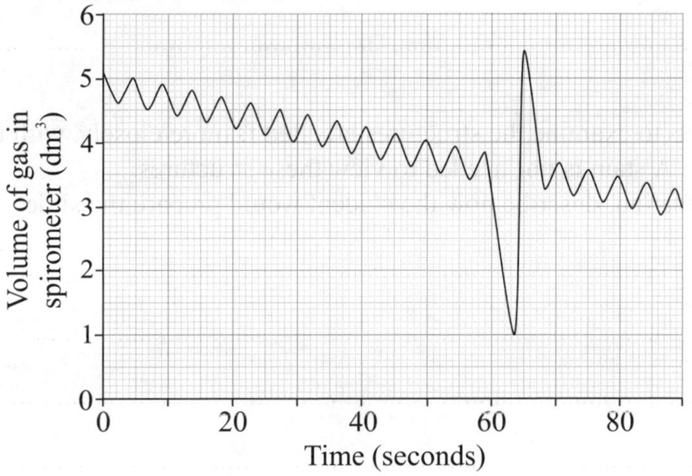

From the graph above, what is this individual's tidal volume?

A 2.1 dm^3 **B** 4.4 dm^3 **C** 0.5 dm^3 **D** 2.9 dm^3

Your answer ☐ **(1 mark)**

3 The following statements describe the process of ventilation in mammals. They are **not** in the correct order.

1 The external intercostal muscles and diaphragm muscles relax.
2 The pressure in the lungs decreases causing air to flow into the lungs.
3 The rib cage moves upwards and outwards, and the diaphragm flattens.
4 The pressure in the lungs increases and air is forced out of the lungs.
5 The external intercostal muscles and diaphragm muscles contract.
6 The rib cage moves downwards and inwards, and the diaphragm becomes curved.

Which option gives the correct order to describe the process of inspiration followed by expiration?

A 5, 3, 2, 1, 6, 4 **B** 1, 6, 4, 5, 3, 2 **C** 5, 6, 4, 1, 3, 2 **D** 1, 3, 2, 5, 6, 4

Your answer ☐ **(1 mark)**

4 A student dissected a grasshopper. As part of the dissection, she removed a piece of the grasshopper's exoskeleton.

(a) Suggest a tool that the student could have used to cut through the exoskeleton.

....

(1 mark)

(b) **Figure 1** shows a diagram of the grasshopper's gas exchange system.

Figure 1

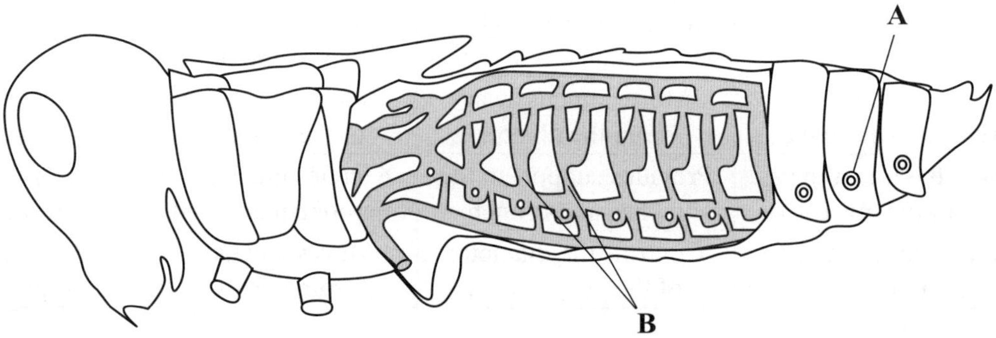

(i) Identify the structures labelled **A** and **B** in **Figure 1**.

A...

B ....

(2 marks)

(ii) The student wants to examine the structures labelled **B** more closely, with the use of a wet mount slide. A stain is **not** needed to view these structures.
Using this information, describe how the student would prepare the slide.

....

....

....

....

(3 marks)

(c) Explain how a grasshopper pumps air in and out of its body.

....

....

....

(2 marks)

(d) The grasshopper's gas exchange system contains fluid.
Describe how this fluid is involved in gas exchange.

....

....

....

(2 marks)

5 Figure 2 shows a gill filament of a bony fish.

Figure 2

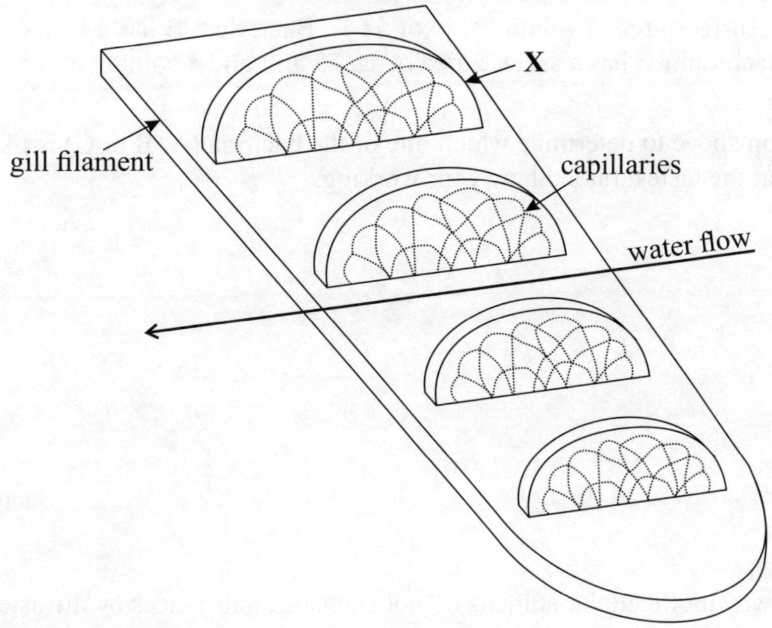

(a) (i) Name structure **X** on **Figure 2.**

....

(1 mark)

(ii) Draw an arrow on structure **X** to show the direction of blood flow.

(1 mark)

(iii) Label structure **X** to show where the highest and lowest concentrations of oxygen are found in the blood.

(1 mark)

(b) Explain **one** way in which the structure of the gill filament is adapted to its function.

....

....

(2 marks)

(c) Explain how changes in the buccal cavity of a bony fish allow the gills to be ventilated.

....

....

....

....

....

(4 marks)

EXAM TIP

Don't panic if you haven't carried out a practical task that you get asked about in the exams. You'll know the skills involved, even if you didn't use them for the same investigation. For example, you might have dissected the gas exchange system of a bony fish rather than an insect, but the principles of dissection and examining microscope slides are the same.

Score

22

Exchange and Transport — 2

1 Bacterium **A** has a surface area to volume ratio of 5 : 1. Bacterium **B** has a surface area to volume ratio of 84 : 22. Bacterium **C** has a surface area of 15.75 μm² and a volume of 3.5 μm³.

(a) Use the information above to determine which **one** of the bacteria (**A**, **B** or **C**) is likely to be able to carry out gas exchange at the fastest rate. Show your working.

Bacterium: ....

(2 marks)

(b) Give **two** reasons why multicellular animals do not exchange substances by diffusion across their outer surface.

1. ....

....

2. ....

....

(2 marks)

2 Sharks exchange oxygen across their gill plates using a counter-current gas exchange system. **Figure 1** shows how the relative oxygen concentration of water changes with distance along a shark's gill plate.

Figure 1

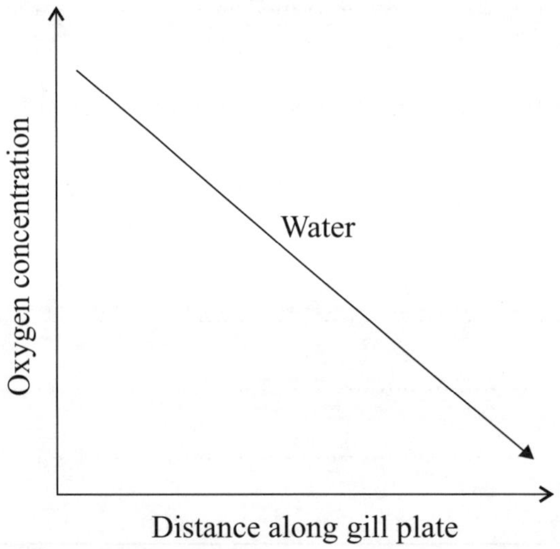

(a) On **Figure 1**, sketch the relative oxygen concentration of the blood flowing through the gill plate.

(1 mark)

Figure 2 shows how the relative oxygen concentrations of water and blood would change with distance along a shark's gill plate if gas exchange took place via a parallel flow system.

Figure 2

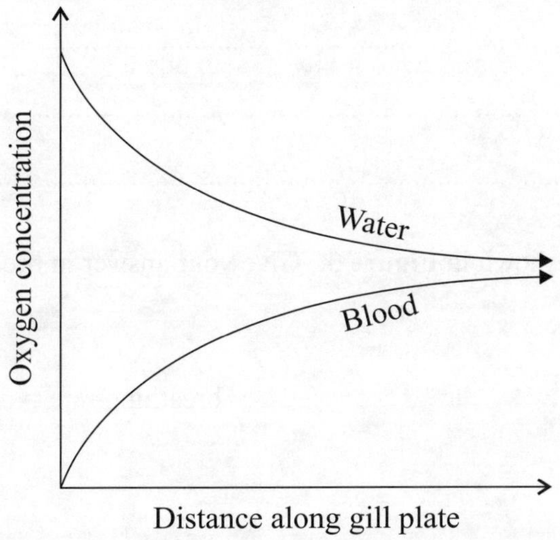

(b) Use **Figure 2** to explain why a parallel flow gas exchange system would be less efficient than a counter-current gas exchange system.

....

....

....

....

....

(3 marks)

3 Intrapulmonary pressure is the pressure inside the lungs.
Figure 3 shows how intrapulmonary pressure changes during breathing.

Figure 3

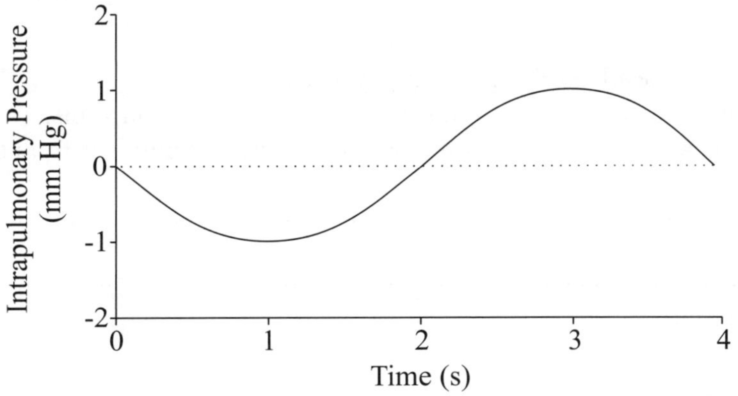

(a) State the time in **Figure 3** at which the lung volume is at its smallest.

....

(1 mark)

(b) State the time period in **Figure 3** during which air is being taken into the lungs.
Explain your answer.

....

....

....

....

(3 marks)

(c) Calculate the breathing rate shown in **Figure 3**. Give your answer in breaths min^{-1}.

breathing rate =.................................... breaths min^{-1}

(1 mark)

4 Chronic bronchitis is a long-term infection of the bronchi. This causes the bronchi walls to become inflamed (swollen) and produce more mucus than usual. This results in a cough and can cause shortness of breath.

(a) Suggest and explain why chronic bronchitis may result in shortness of breath.

....

....

....

....

(3 marks)

(b) Explain how the concentration gradients of oxygen and carbon dioxide are maintained between the alveoli and the blood in a healthy person.

....

....

....

(2 marks)

(c) Chronic bronchitis often develops alongside emphysema.
Emphysema is a lung disease that leads to the breakdown of the alveoli walls.
Suggest why a person with emphysema would have difficulty expelling air from their lungs.

....

....

....

(2 marks)

EXAM TIP

Ah, 'context'... the examiners just love it. That's why, instead of asking you bog standard questions about diffusion or ventilation in the lungs, they're more likely to get you to apply your knowledge to a context — for example, by telling you all about a disease that interrupts lung function and then getting you to work out why that's the case. If you've learnt your stuff, you'll be fine.

Score

20

Transport in Animals — 1

More questions about transport now — this time they're all about circulatory systems in animals. If your heart rate is quickening in panic at the thought of this stuff coming up in the exams, you better get practising now.

For each of questions 1-3, give your answer by writing the correct letter in the box.

1 Look at the ECG below.

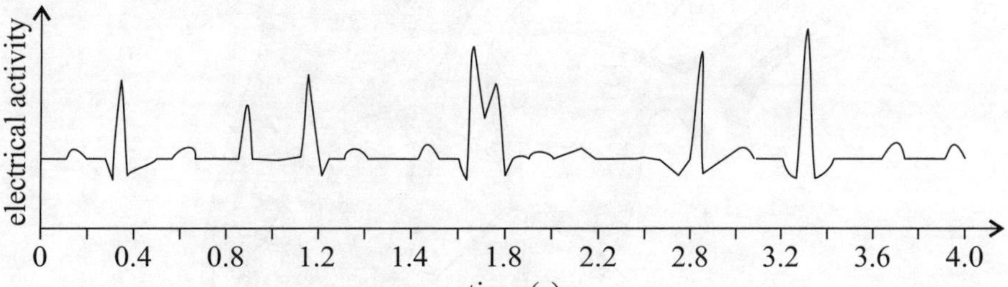

What abnormal heart activity is displayed in this ECG?

A fibrillation

B an ectopic heartbeat

C tachycardia

D bradycardia

Your answer ☐ **(1 mark)**

2 Which of the following statements related to the function of haemoglobin is/are correct?

Statement 1: At low partial pressures of oxygen, oxygen is loaded onto haemoglobin, forming oxyhaemoglobin.

Statement 2: Oxyhaemoglobin more readily dissociates into haemoglobin and oxygen when the concentration of carbon dioxide in the blood is high.

Statement 3: Fetal haemoglobin has a higher affinity for oxygen than adult haemoglobin.

A 1, 2 and 3

B 1 and 2 only

C 2 and 3 only

D 1 only

Your answer ☐ **(1 mark)**

3 During the cardiac cycle, blood enters the heart through the vena cava and pulmonary vein. The following statements describe the rest of the cardiac cycle. They are **not** in the correct order.

1: The atria contract, forcing blood through the atrioventricular valves.

2: The wave of electrical activity is detected by the atrioventricular node and passed along the bundle of His and the Purkyne tissue.

3: The ventricles contract, forcing blood through the semi-lunar valves.

4: The sino-atrial node sends out a wave of electrical activity.

5: Blood leaves the heart through the aorta and pulmonary artery.

Which option gives the correct order of events in rest of the cardiac cycle?

A 1, 4, 2, 3, 5

B 4, 3, 2, 1, 5

C 3, 5, 4, 2, 1

D 4, 1, 2, 3, 5

Your answer ☐ **(1 mark)**

4 Electrical impulses in the heart are generated by the sino-atrial node (SAN) and conducted across the muscle tissue in a highly coordinated sequence.

Figure 1 shows a diagram of the heart. The SAN and atrio-ventricular node (AVN) are labelled.

Figure 1

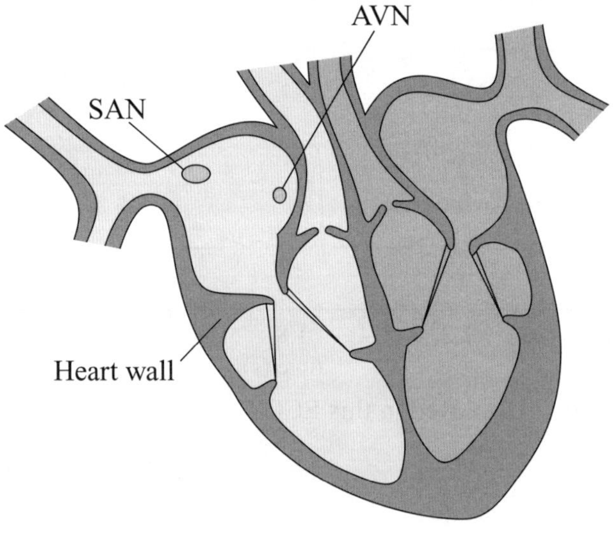

(a) Describe the function of the AVN.

....

(1 mark)

The heart wall contains muscle tissue and collagen, which acts as a structural support. Collagen does not conduct electricity.

(b) Explain why the collagen in the heart wall also has a role in helping to ensure that the ventricles contract at the right time.

....

....

....

(2 marks)

5 **Figure 2** shows a cross-section of the human heart.

Figure 2

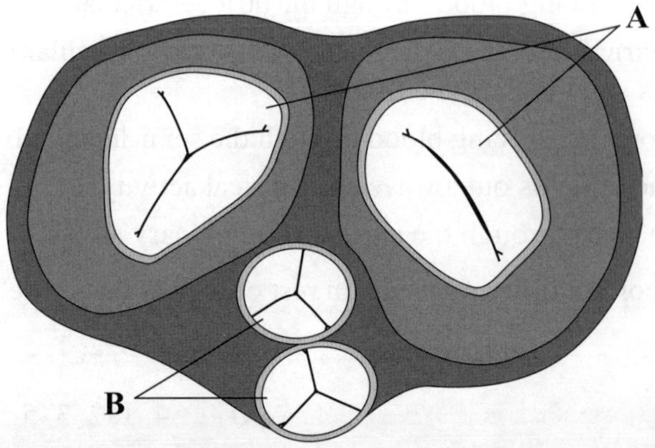

(a) What is the role of the valves labelled **A** in **Figure 2**?

....

....

....

(2 marks)

(b) In terms of pressure changes in the heart, explain what causes the valves labelled **B** in **Figure 2** to open.

....

....

(1 mark)

(c) Cardiac output is the volume of blood pumped out of the left ventricle in one minute. Scientists investigated the effect of body position on heart rate and cardiac output. **Table 1** shows their results.

Table 1

	Standing up	Lying down
Mean heart rate (bpm)	74	57
Mean cardiac output (cm^3 min^{-1})	4700	4700
Mean stroke volume (cm^3)		

(i) The stroke volume is the volume of blood that is pumped out by the left ventricle in one cardiac cycle. Use the information in **Table 1** to complete the table to show the mean stroke volumes. Give your answers to three significant figures. Show your working.

(2 marks)

(ii) The scientists ensured that the participants of the investigation had been in the required position for five minutes before they recorded these measurements. Suggest why.

....

....

(1 mark)

(iii) Suggest why there is a difference in heart rate between standing up and lying down in **Table 1**.

....

....

(1 mark)

(iv) Explain why the scientists used 'mean' measurements.

....

....

(1 mark)

EXAM TIP

Try not to leave anything blank in the exam, especially multiple choice questions. You won't lose marks for wrong answers, so you might as well try narrowing down your options, then give your best guess. It can be easier than it looks to rule things out — e.g. if you have to choose the right sequence of events, think about whether certain things must happen before others.

Score

14

Transport in Animals — 2

1 A student carried out a dissection of a mammalian heart.
Figure 1 shows the external structure of the heart.

Figure 1

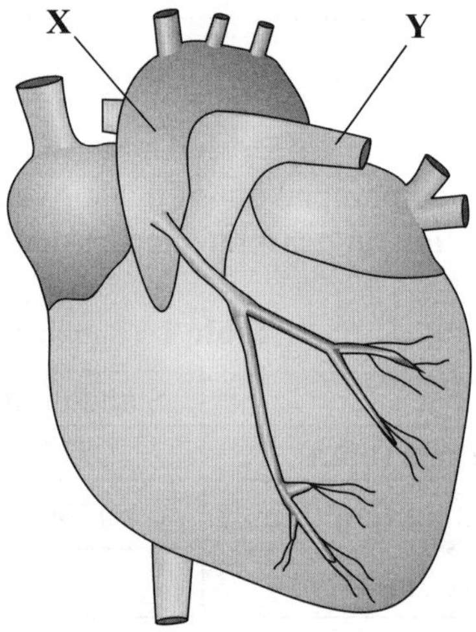

(a) (i) Name the blood vessels labelled **X** and **Y** on **Figure 1**.

X: ....

Y: ....

(2 marks)

The student struggled to identify the blood vessels attached to the heart.
He was also unable to tell which were arteries and which were veins.

(ii) Describe how feeling the inside of the blood vessels would have allowed the students to tell the difference between an artery and a vein.

....

....

(1 mark)

(b) Sketch a line on **Figure 1** to show where the student would need to make a cut to see inside the left ventricle.

(1 mark)

(c) Name the blood vessel that can be seen on the surface of the heart in **Figure 1**.

....

(1 mark)

The student produced a biological drawing of the heart.

(d) Give **one** instruction that the student would need to follow in order to produce a clear and useful drawing.

....

(1 mark)

2 Lymph is formed from tissue fluid, which is formed in capillary beds.

(a) Lymphoedema is a condition that develops when a person's lymphatic system does not function correctly. The main symptom is the swelling of parts of the body.

(i) Suggest why lymphoedema causes swelling.

....

....

(1 mark)

(ii) State **two** ways in which the composition of lymph is different from the composition of blood.

1. ....

2. ....

(2 marks)

(b) **Table 1** shows the hydrostatic and oncotic pressures along a capillary bed.

Table 1

	Arteriole end of capillary bed	Venule end of capillary bed
Hydrostatic pressure (mmHg)	33	16
Oncotic pressure (mmHg)	26	26

(i) Explain the difference in hydrostatic pressure at the arteriole and venule ends of the capillary bed.

....

....

(2 marks)

(ii) Suggest why the oncotic pressure remains the same at each end of the capillary bed.

....

....

....

(2 marks)

(c) Capillary beds are important exchange surfaces.
Explain **one** way in which the structure of capillaries helps them carry out their function.

....

....

(1 mark)

3 **Figure 2** shows the oxygen dissociation curve for a person who is exercising, alongside a person who is not exercising.

Figure 2

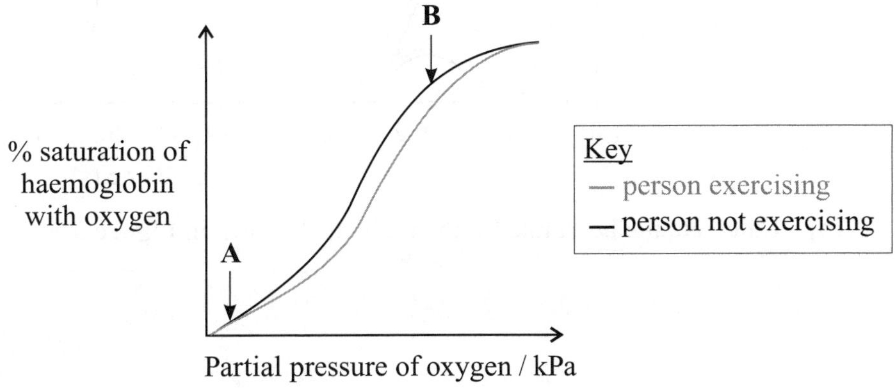

(a) Explain why the percentage saturation of haemoglobin with oxygen is lower at point **A** in **Figure 2** than at point **B**.

....

....

....

(2 marks)

(b) (i) Explain why the Bohr effect means that the oxygen dissociation curve for the person exercising is to the right of the oxygen dissociation curve for the person who is not exercising.

....

....

....

(2 marks)

(ii) Describe the roles of carbonic anhydrase and carbonic acid in the Bohr effect.

....

....

....

....

(3 marks)

4 **Figure 3** shows the pressure changes in an individual's heart during the cardiac cycle.

Figure 3

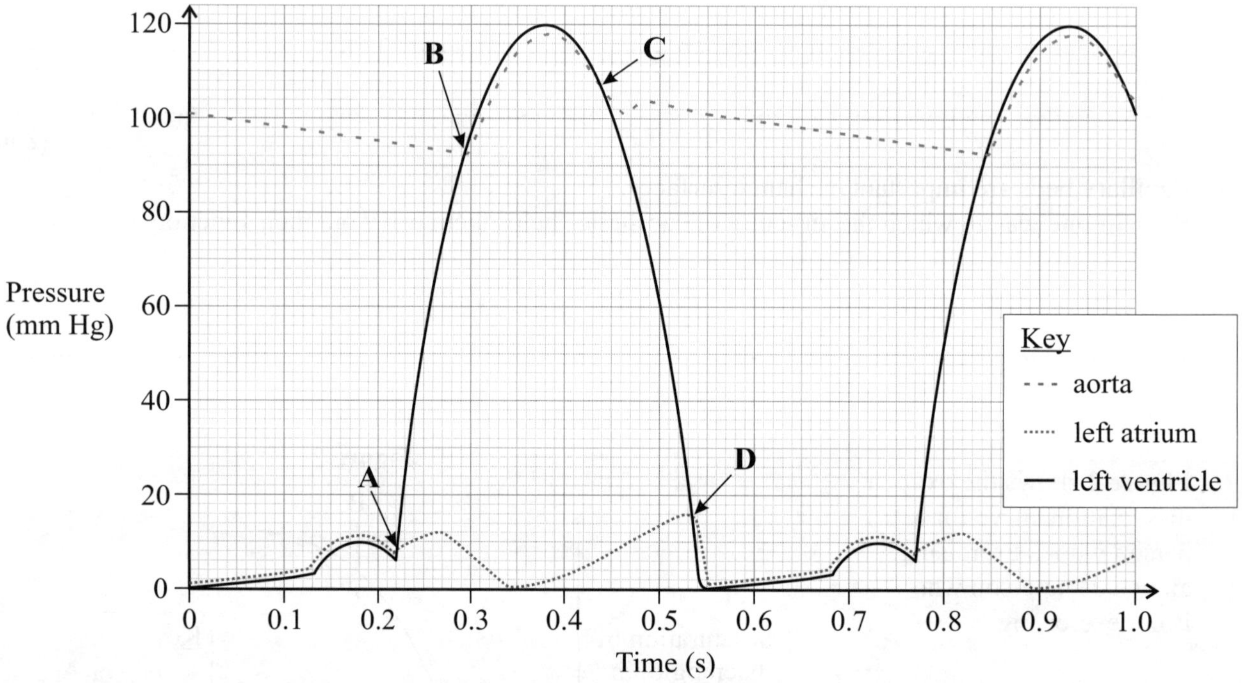

(a) Calculate the heart rate for the individual shown in **Figure 3**.

.... beats per minute

(1 mark)

(b) When on **Figure 3** does the left atrium first start to contract?

Hint: when a chamber contracts, there's a sudden rise in pressure inside the chamber.

.... seconds

(1 mark)

(c)* Explain the events that are occurring at points **A** to **D** on **Figure 3**.

....

....

....

....

....

....

....

....

....

....

....

....

....

(6 marks)

(d) Explain the difference between the maximum pressures of the left atrium and the left ventricle of the heart.

....

....

....

(1 mark)

(e) Using your own knowledge and information from **Figure 3**, explain **one** way in which the aorta is adapted for its function.

....

....

....

(2 marks)

EXAM TIP

Don't worry if you're presented with an unfamiliar graph in the exam. Just make sure you read the key carefully (if there is one) and the labels on the axes too. Don't try to answer any questions until you're sure you understand exactly what the graph is showing.

Score

32

Transport in Plants

Now's the time to find out whether you can remember everything you should know about transport in plants, or whether you've got a memory like a sieve tube element.

For each of questions 1-3, give your answer by writing the correct letter in the box.

1 The rate of transpiration in plants is affected by a variety of factors, including light intensity, temperature, humidity and wind speed. In the table below, which of the rows **A** to **D** corresponds to the set of conditions that would give the fastest rate of transpiration?

	Light intensity	Temperature	Humidity	Wind speed
A	low	high	low	low
B	high	high	low	high
C	high	high	high	high
D	low	low	low	low

Your answer ☐

(1 mark)

2 Which of the statements, **A** to **D**, related to the transport of water in plants, is **incorrect**?

A Water travels down a water potential gradient to enter a plant through the root hair cells.

B Water must travel from the root hair cells through the root cortex to reach the xylem.

C When water is transported through areas of the plant via the apoplast pathway, it is transported through the non-living parts of plant cells, such as the cell walls.

D Water molecules moving through the apoplast pathway pass through plasmodesmata.

Your answer ☐

(1 mark)

3 Which of the following statements about adaptations of hydrophytes is/are **true**?

Statement 1: Hydrophytes have stomata in sunken pits and a thick waxy layer on their epidermis.

Statement 2: Stomata are usually only found on the upper surface of hydrophytes' leaves.

Statement 3: Air spaces in the tissues of hydrophytes can act as a store of oxygen for use in respiration.

A 1, 2 and 3

B 1 and 2 only

C 2 and 3 only

D 3 only

Your answer ☐

(1 mark)

4 A student used a potometer to investigate transpiration in a plant. **Figure 1** shows how the potometer was set up. The student closed the tap, then took the capillary tube out of the beaker of water long enough for an air bubble to form. She then recorded the amount of time it took for the air bubble to move between the two markers, and used it to calculate the transpiration rate.

Figure 1

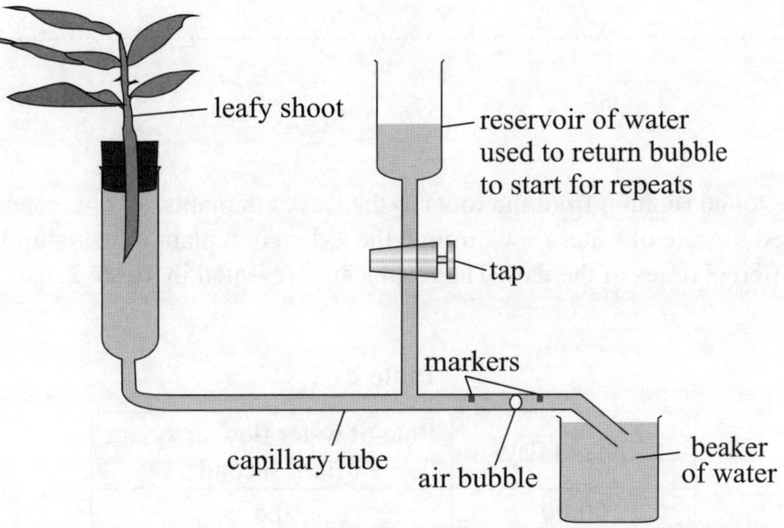

(a) When setting up this experiment, it is important that water does not touch the leaves. Using your knowledge of water transport in plants, explain why.

....

....

....

(2 marks)

(b) Suggest **one** other precaution that must be taken when setting up this experiment.

....

(1 mark)

(c) Suggest and explain why measuring the time it takes for the air bubble to move between the two markers may not result in an accurate estimation of the plant's transpiration rate.

....

....

....

(2 marks)

The student used the potometer to investigate the effect of temperature on the transpiration rate. The results are shown in **Table 1**.

Table 1

Temperature / °C	Mean transpiration rate / cm^3 per minute
20	0.20
30	0.34
40	0.85
50	1.24
60	1.36

(d) Explain the relationship between the temperature and the transpiration rate in **Table 1**.

....

....

....

....

(3 marks)

5 Xylem vessels are found running from the roots to the leaves in plants. Scientists measured the rate of water flow through the xylem of a plant in its natural environment at different times of the day. The results are presented in **Table 2**.

Table 2

Time of day	Rate of water flow in xylem / mm^3 $second^{-1}$
00:00	0.8
06:00	2.8
12:00	4.5
18:00	3.0

(a) Using your knowledge of water transport in the xylem, explain the difference in the results between 00:00 and 12:00 in **Table 2**.

....

....

....

....

(3 marks)

(b) In order to investigate the structure of the xylem, the scientists also carried out a dissection of the plant's stem.

(i) Suggest why the scientists kept the dissected plant tissue in water until they were ready to view the cells.

....

....

(1 mark)

(ii) Suggest a further step that the scientists would have carried out in order to observe the xylem vessels under a microscope.

....

....

(1 mark)

(iii) Describe **one** way in which the position of xylem vessels differs between the stem and the root of a dicotyledonous plant.

....

....

(1 mark)

6 Translocation in a plant describes the movement of solutes to where they are needed within the plant.

(a) When treated with metabolic inhibitors, translocation in a plant stops.
Explain why.

....

....

....

(2 marks)

Figure 2 represents the flow of solutes in the phloem of a tree, according to the mass flow hypothesis.

Figure 2

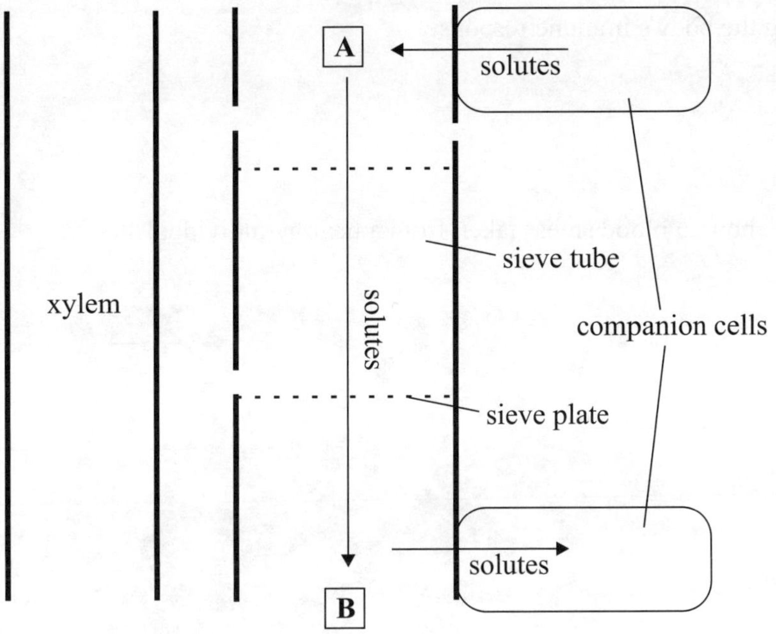

(b) Explain why the pressure at point **A** in **Figure 2** will be higher than that at point **B**.

....

....

....

....

....

(4 marks)

EXAM TIP

Make sure you pay attention to whether a question is asking you to 'describe' or 'explain' — they don't mean the same thing. 'Describing' means giving an account of something, e.g. saying how a variable changes in a table or graph, whereas 'explaining' means setting out reasons behind it, e.g. saying why the variable changes.

Score

23

Disease and the Immune System — 1

These questions are about disease and the immune system. Your primary response is going to want you to turn away from this page, but make sure you stick at it. You want to get those memory cells activated for your exam.

For questions 1 and 2, give your answer by writing the correct letter in the box.

1 Rheumatoid arthritis is an autoimmune disease that causes inflammation of the joints. Which of the following statements, **A** to **D**, related to rheumatoid arthritis is **true**?

A Rheumatoid arthritis can be caused by the body failing to produce antibodies against foreign antigens in the joints.

B Rheumatoid arthritis can be caused by the body mistakenly producing antibodies against antigens present on cells in the joints.

C Rheumatoid arthritis can be caused by the body producing excess antibodies in response to the presence of foreign antigens in the joints.

D Rheumatoid arthritis can be caused by antigens present on cells in the joints suppressing the body's immune response.

Your answer ☐

(1 mark)

2 The image below shows a blood smear taken from a healthy individual.

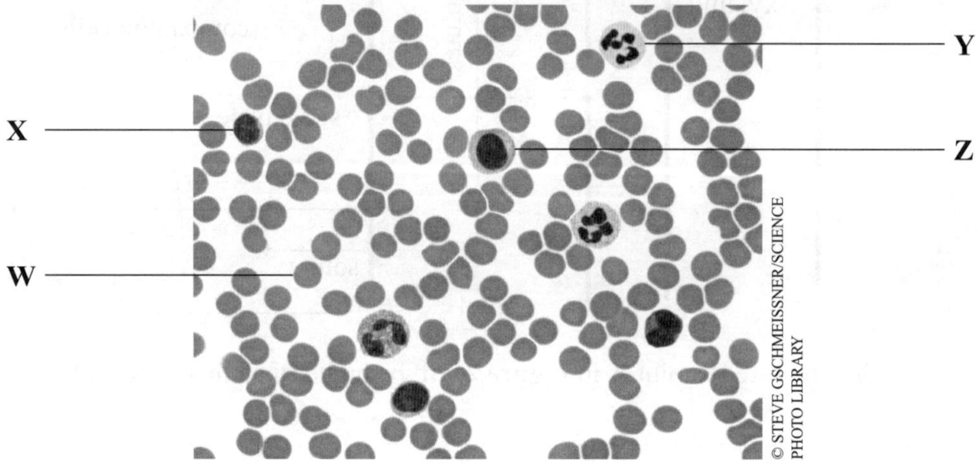

Which of the rows, **A** to **D**, in the table below correctly identifies the cells labelled **W** to **Z**?

	W	X	Y	Z
A	Lymphocyte	Monocyte	Neutrophil	Red blood cell
B	Red blood cell	Monocyte	Lymphocyte	Neutrophil
C	Lymphocyte	Neutrophil	Monocyte	Red blood cell
D	Red blood cell	Lymphocyte	Neutrophil	Monocyte

Your answer ☐

(1 mark)

3 **Figure 1** shows an antibody. The antibody acts as an agglutinin.

Figure 1

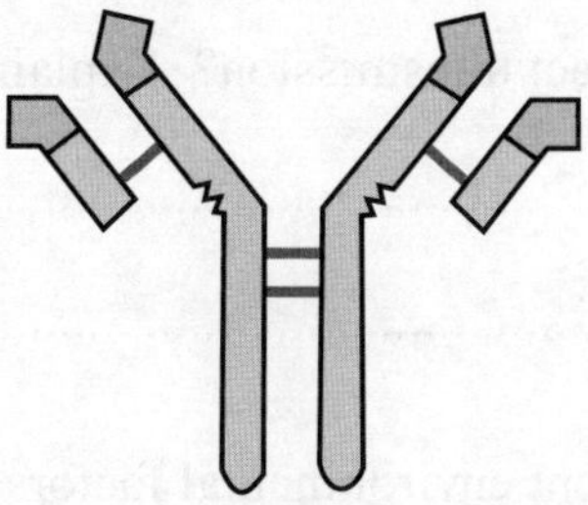

(a) (i) Describe the role of this antibody in the immune response.

....

(1 mark)

(ii) Name a type of cell that produces antibodies.

....

(1 mark)

(iii) Describe **two** ways in which the structure of the antibody shown in **Figure 1** is adapted to its function.

1. ....

2. ....

(2 marks)

Some antibodies have a more complex structure, made up of several monomers joined together. This is shown in **Figure 2**.

Figure 2

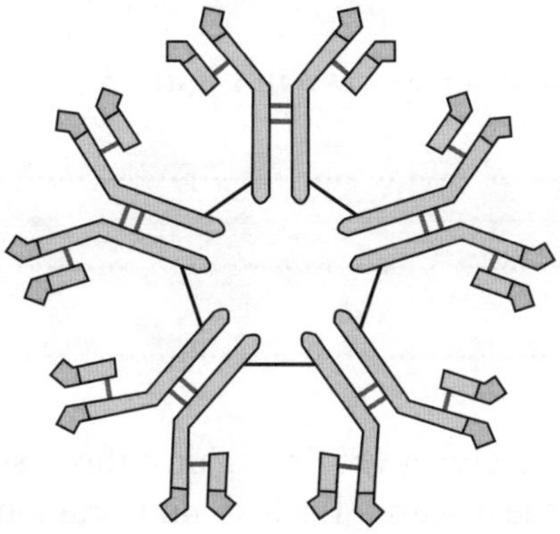

(b) Suggest and explain **one** advantage of the structure in **Figure 2**, compared to that in **Figure 1**.

....

....

(2 marks)

(c) Some cells can produce antibodies at a rate of 2000 molecules per second.
Calculate how many antibodies would be produced by one of these cells in one hour.
Show your working. Give your answer in standard form.

Number of antibodies = ...

(2 marks)

4 Scientists are investigating the transmission of a diarrhoeal disease that can be fatal in young children. The disease can be contracted by drinking water contaminated with infected faeces. For this reason, it is common in low-income countries with inadequate sewage and drinking water systems.

(a) Does this disease show direct or indirect transmission? Explain your answer.

....

....

(1 mark)

The scientists investigated how different environmental factors influenced the spread of this disease in a particular low-income country. **Figure 3** shows some of their results.

Figure 3

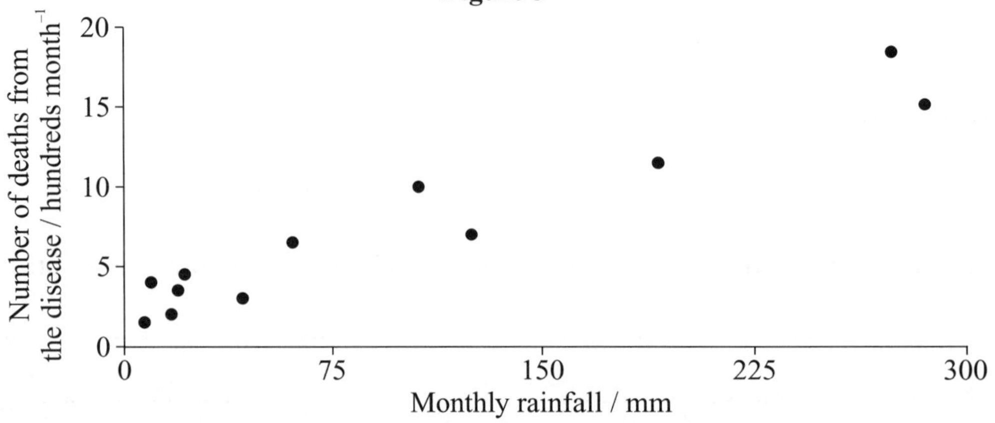

(b) (i) What conclusion can be drawn from the results shown in **Figure 3**?

....

....

(1 mark)

(ii) Suggest an explanation for the results shown in **Figure 3**.

....

....

....

(2 marks)

(c) Other scientists are attempting to develop a vaccine against this disease. There are several different strains of the pathogen that cause the disease. Each strain is immunologically distinct. Suggest and explain why this may make it difficult to develop an effective vaccine against the disease.

....

....

....

....

(3 marks)

EXAM TIP

Make sure you understand what a graph is showing before you launch into your answer. You should ~~double~~ triple check the axis labels — you don't want to talk about the number of cases of the disease, for example, if the graph's talking about the number of deaths from the disease.

Score

17

Disease and the Immune System — 2

1 Corn plants that are being eaten by caterpillars release a chemical. The chemical's odour attracts a species of wasp that lays its eggs in the caterpillars. When the eggs hatch, the caterpillars are killed.

(a) Suggest **one** way in which a corn plant's response to being eaten may help to defend it against infection by pathogens.

....

....

(1 mark)

(b) Describe **one** way in which a corn plant may be able to limit the spread of an invading pathogen between its cells.

....

....

(1 mark)

2 Phagocytosis is an important step in the immune response of humans.

(a) When a person is wounded, phagocytes migrate from different areas of the body to the site of the wound.

(i) Explain what causes phagocytes to migrate in this way.

....

....

(2 marks)

(ii) Suggest a benefit of the phagocytes' migration response.

....

....

(1 mark)

(b) Describe what happens when a phagocyte comes into contact with a pathogen.

....

....

....

....

(3 marks)

IgG is an opsonin. Some people do not produce enough IgG.

(c) Explain how phagocytosis may be affected by a reduced level of IgG in the blood.

....

....

....

(2 marks)

3 **Figure 1** shows a person's immune responses following two separate infections with the same strain of bacteria.

Figure 1

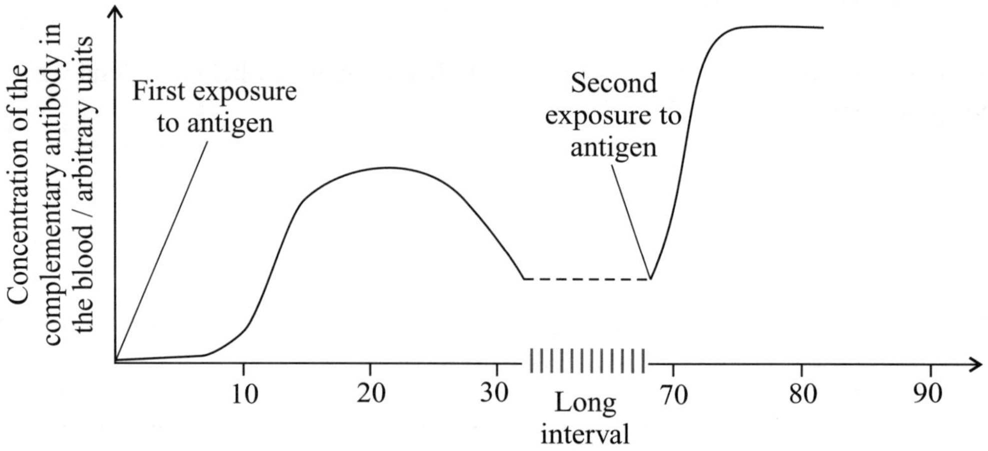

(a) (i) Explain the shape of the curve for the primary immune response.

Remember, 'explain' means 'give reasons for'. Don't just describe the shape of the curve here.

....

....

....

....

....

....

(4 marks)

(ii) Using the information shown in **Figure 1**, explain why the person shows no symptoms following the second exposure to a bacterial antigen.

....

....

....

(2 marks)

Tetanus is a disease caused by *Clostridium tetani* bacteria. The bacteria secrete a toxin that can cause serious muscle spasms. Tetanus can be treated with a mixture of anti-toxins and antibiotics.

(b) Explain why anti-toxins are an effective treatment for tetanus but cannot be used to cure the disease.

....

....

....

....

(3 marks)

4 Group B streptococcus (GBS) is a bacterium that can be carried, without harm, in a healthy human body. However, if a newborn baby becomes infected with GBS, it can lead to meningitis or other serious diseases.

GBS is typically passed on from a carrier mother to her baby during birth. Scientists are currently working to develop a vaccine against GBS, which will be given to pregnant women.

(a) (i) Suggest why a pregnant woman may be vaccinated against GBS even if she carries it without harm.

....

....

(1 mark)

(ii) Explain how the vaccine would lead to the production of memory cells against GBS.

....

....

....

....

....

(4 marks)

When a baby breastfeeds, it receives some of its mother's antibodies.
This gives the baby immunity against the diseases its mother is immune to.

(b) Describe **two** differences between the immunity obtained from breastfeeding with the immunity obtained from a vaccine.

1. ....

....

2. ....

....

(2 marks)

Meningitis is the inflammation of the protective layers around the brain and spinal cord.

(c) (i) Using the information provided, explain how the non-specific immune response to a GBS infection could lead to meningitis.

Hint: this question is really asking you how the non-specific immune response against bacteria could lead to inflammation.

....

....

....

....

(3 marks)

(ii) Meningitis can be treated with antibiotics.
Explain why it might become a problem to treat meningitis with antibiotics in future.

....

....

(1 mark)

5 **Figure 2** shows a simplified model of the different antigens present on red blood cells from different blood types.

Figure 2

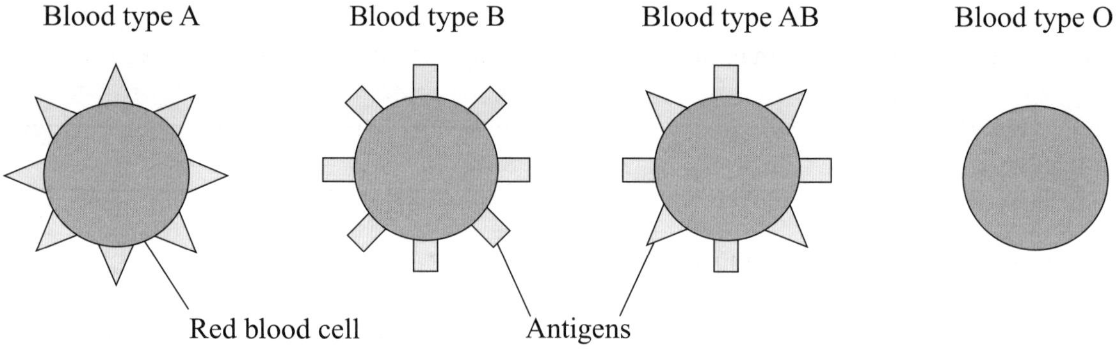

(a)* Using the information in **Figure 2**, explain what would happen if a person with blood type A received a transfusion of blood type B.

....

....

....

....

....

....

....

....

....

....

....

(6 marks)

(b) Explain why anyone can receive type O blood.

....

....

(1 mark)

(c) Agglutinins can be used to determine whether a person has type O blood. Suggest how.

....

....

(2 marks)

EXAM TIP

When answering exam questions — especially ones like 5 (a)*, where you'll be marked on the quality of your extended response — it's important to use the correct, scientific terms. There are a lot of scientific terms in this topic (e.g. phagocytosis, opsonin and antigen to name just a few) so make sure you learn what they all mean <u>and</u> how to spell them.

Score

39

Biodiversity

There's a lot you need to know about biodiversity — what it is, how to measure it, which factors affect it, and how to conserve it. You know what that means... here's a bunch of questions about all those things. Enjoy.

For each of questions 1-3, give your answer by writing the correct letter in the box.

1 Some students are investigating the species present in a coastal habitat. They draw a straight line from the coast, heading inland, and then use quadrats to sample 1 m^2 areas at regular intervals along the line. What type of sampling, **A** to **D**, are the students using?

A random sampling

B systematic sampling

C stratified sampling

D opportunistic sampling

Your answer ☐

(1 mark)

2 Which of the following statements about *in situ* and *ex situ* conservation methods is/are **true**?

Statement 1: *In situ* conservation might involve preventing the introduction of non-native species that threaten local species.

Statement 2: An advantage of *ex situ* conservation is that it is easier to control the environment than with *in situ* conservation.

Statement 3: Marine conservation zones are an example of *ex situ* conservation.

A 1, 2 and 3

B 1 and 2 only

C 2 and 3 only

D 1 only

Your answer ☐

(1 mark)

3 A scientist is using Simpson's Index of Diversity to compare the level of biodiversity in four different habitats. The results are shown in the table below.

Habitat	Simpson's Index of Diversity
A	0.64
B	0.24
C	0.72
D	0.45

Based on the information above, which of the following statements is/are definitely **true**?

Statement 1: Habitat A has a higher species richness than Habitat D.

Statement 2: Habitat B has the highest biodiversity.

Statement 3: Habitat C is more diverse than Habitat D.

A 1, 2 and 3

B 1 and 3 only

C 3 only

D 2 only

Your answer ☐

(1 mark)

4 Scientists investigated the diversity of plants in an ungrazed field. The data obtained is shown in **Table 1**.

Table 1

Species	Number of individual plants counted in different quadrats					Mean number counted
Rapeseed	24	46	32	28	32	
Common sunflower	1	0	2	1	1	
Common poppy	8	12	6	10	8	
Creeping thistle	13	14	7	15	13	

(a) (i) Complete **Table 1** to show the mean number of each species counted.

(1 mark)

(ii) Using the mean values you added to **Table 1** and the formula provided below, calculate Simpson's Index of Diversity for plants in this field. Show your working.

$$D = 1 - \left(\sum \left(\frac{n}{N}\right)^2\right)$$

where n = total number of individuals of one species
and N = total number of organisms of all species

D = ...

(2 marks)

(b) Suggest a method that the scientists could have used to randomly place the quadrats in the field.

....

....

(1 mark)

The scientists then gathered data in a second field on which farm animals were allowed to graze. The mean number of creeping thistle was much lower in the second field than in the first.

(c) The scientists wanted to determine if the difference in means between the two fields was significant. State which statistical test they could have used to determine this. Explain your choice.

....

....

(2 marks)

(d) Further investigations in both fields showed the species diversity of the grazed field to be lower than that of the ungrazed field. Suggest an explanation for this.

....

....

....

(2 marks)

5 Scientists wanted to investigate the impact of different farming practices on ladybird biodiversity. To do so, they counted the number of different ladybird species on organic and conventional farms. This allowed them to compare the species richness of the ladybirds in the different types of farm.

(a) Explain why it may have been more useful for the scientists to compare indexes of diversity for their investigation.

....

....

....

(2 marks)

The scientists' data can be seen in **Figure 1**. The error bars indicate standard deviation.

Figure 1

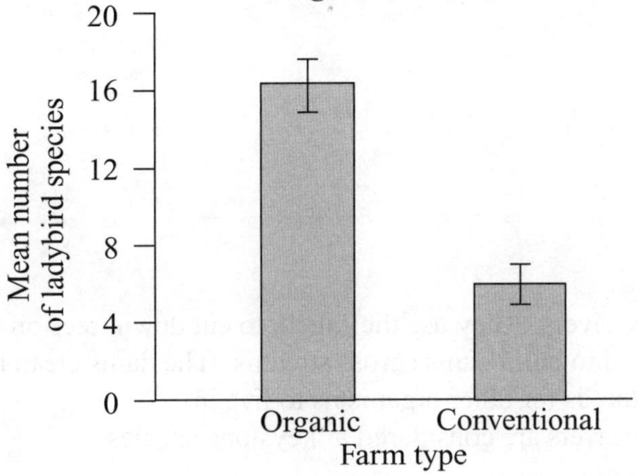

(b) The scientists concluded that conventional farming had a much greater impact on the number of ladybird species than organic farming. Use the data in **Figure 1** to evaluate this claim.

....

....

....

....

....

(3 marks)

Many of the conventional farms studied by the scientists practised continuous monoculture.

(c) (i) Explain **one** reason why continuous monoculture has a negative impact on biodiversity.

....

....

(1 mark)

(ii) Explain **one** economic reason **against** growing crops using continuous monoculture.

....

....

....

(2 marks)

6 A captive breeding programme is being carried out with beavers. Genetic samples are taken from the beavers, so that the proportion of polymorphic gene loci can be monitored.

(a) (i) Suggest why it is important to monitor the proportion of polymorphic gene loci in this population of beavers.

(2 marks)

(ii) 165 loci in the genome of the beaver population are sampled.
132 of the gene loci have only one allele present, while the rest have more than one allele.
Calculate the proportion of the gene loci that are polymorphic. Show your working.

(2 marks)

(b) Beavers live in and around rivers. They use their teeth to cut down trees on the river bank, which they use for food and to build dams across streams. The dams clean the water and create new ponds and wetlands for other organisms to live in.
Suggest **one** reason why beavers are considered a 'keystone species'.

(1 mark)

(c) It is hoped that members of the captive beaver population could eventually be released into wild. This is part of an attempt to increase the biodiversity of their natural river habitat.
Suggest **one** way in which human population growth may have reduced the biodiversity of this habitat.

(1 mark)

(d) Conservation agreements can help to improve biodiversity.
Describe how **one** international or local conservation agreement protects species or habitats.

(2 marks)

EXAM TIP

A maths question might seem simple if you're given the formula, but it's still easy to make mistakes. You need to make sure that you're substituting the correct numbers into the formula. For example, with the formula for Simpson's Index of Diversity, make sure you don't get 'N' (number of all organisms) and 'n' (number of individuals of one species) mixed up.

Score

27

Classification and Evolution — 1

You might not classify this section among the most fun things you could be doing right now, but you really ought to have a go at these questions about classification and evolution. It'll be worth it in the end.

For questions 1 and 2, give your answer by writing the correct letter in the box.

1 The statements below describe the features of four different types of single-celled organism.

Blepharisma japonicum: A protozoan with a nucleus that occurs in water or soil.
Nitrospira moscoviensis: An organism with no nucleus and a helical shape that can oxidise nitrites.
Stipitococcus capensis: A type of algae found in freshwater that contains chlorophyll and is capable of photosynthesis.
Brettanomyces bruxellensis: A saprotrophic organism with a cell wall made of chitin commonly used in the production of beer.

Which row, **A** to **D**, of the table below gives the correct classification for each of these organisms according to the five kingdom classification system?

	Blepharisma japonicum	*Nitrospira moscoviensis*	*Stipitococcus capensis*	*Brettanomyces bruxellensis*
A	Fungi	Protoctista	Protoctista	Fungi
B	Protoctista	Prokaryotae	Protoctista	Fungi
C	Prokaryotae	Fungi	Plantae	Animalia
D	Fungi	Prokaryotae	Plantae	Protoctista

Your answer ☐ **(1 mark)**

2 The diagram below shows a simple phylogenetic tree for a number of types of reptile.

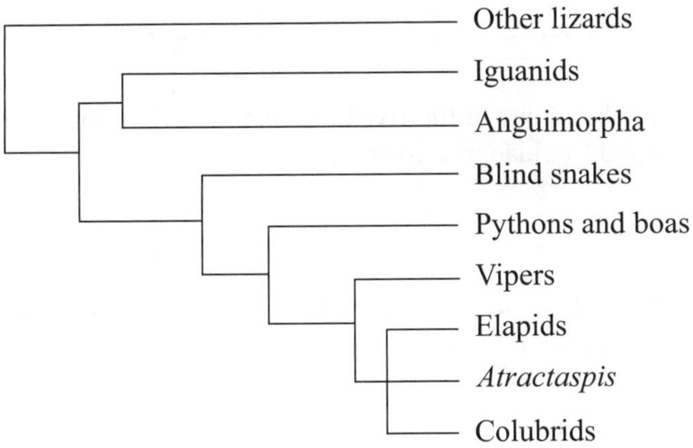

On the basis of this diagram, which of the following statements is/are true?

Statement 1: Iguanids are more closely related to Anguimorpha than other lizards.

Statement 2: Blind snakes are more closely related to pythons and boas than vipers.

Statement 3: Elapids are more closely related to *Atractaspis* species than colubrids.

A 1, 2 and 3

B 1 and 2 only

C 2 and 3 only

D 1 only

Your answer ☐ **(1 mark)**

3 **Figure 1** is a phylogenetic tree. It shows how different species from the order Carnivora are related.

Figure 1

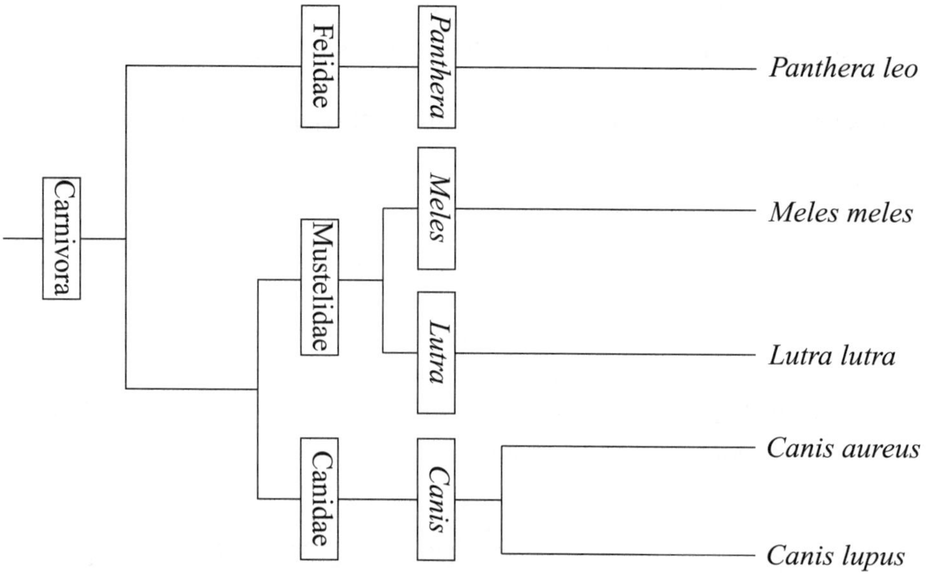

(a) The plural of genus is genera.
How many different genera are represented in **Figure 1**?

....

(1 mark)

(b) Which **two** species in **Figure 1** are most closely related? Give a reason for your answer.

....

....

(1 mark)

(c) Which level of the taxonomic hierarchy of the five kingdom classification system is represented by the groups Felidae, Mustelidae and Canidae in **Figure 1**?

....

(1 mark)

(d) The phylogenetic tree in **Figure 1** is within the domain of Eukarya in the three domain classification system. Describe how molecular evidence led to the five kingdom classification system being replaced by the three domain classification system.

....

....

....

....

(2 marks)

EXAM TIP

There are quite a few terms you need to know for this topic, so make sure you learn them all. Most of the five kingdoms are easy to remember, but try not to get Prokaryotae and Protoctista confused — even though they're similar words and are both made up of fairly simple organisms, there's a crucial difference in that Protoctista are eukaryotic.

Score

7

Classification and Evolution — 2

1 The golden spiny mouse is a species of small mammal found in the Middle East, often living in hot desert habitats. It has a high water intake through its diet, which includes snails, insects and green vegetation. It is also able to make highly concentrated urine. It has spiny fur, which it can spread out when threatened, so that it appears larger than it is.

(a) Describe how **one** physiological adaptation increases the chance of survival in golden spiny mice.

....

....

(1 mark)

(b) The fur of golden spiny mice is a golden-brown colour, which is a similar colour to the ground in the desert habitats that they usually live in.
Explain how this fur colour may have evolved through natural selection.

....

....

....

....

....

....

(4 marks)

(c) Golden spiny mice are also known as *Acomys russatus*.
Explain why there is this other name for the species.

....

....

....

(2 marks)

(d) Five adult golden spiny mice were captured from two populations in different areas and weighed. The results are shown in **Table 1**.

Table 1

	Mass of mouse / g				
Population 1	43.9	39.9	42.6	43.1	37.4
Population 2	44.8	39.8	51.5	50.2	46.0

(i) Two of the individuals from the Population 2 sample are over 50 g.
Explain the potential causes for this variation in mass.

....

....

....

(2 marks)

(ii) Calculate the standard deviation for the results from Population 1, using the formula:

$$s = \sqrt{\frac{\sum (x - \overline{x})^2}{n - 1}}$$

Give your answer to three significant figures.

....

(2 marks)

(iii) Suggest how it could be determined whether there is a significant difference between the two populations shown in **Table 1**.

....

(1 mark)

2 The theory of evolution by natural selection is attributed to Charles Darwin. However, Alfred Russel Wallace also contributed to its formulation.

(a) Describe Wallace's contribution to the formulation of the theory of evolution by natural selection.

....

....

....

(2 marks)

Since Darwin and Wallace's work was first published, more evidence for the theory of evolution by natural selection has been discovered. For example, some fossils of an extinct dinosaur species called *Archaeopteryx* have been found. These fossils had some features of land dinosaurs, such as long, bony tails and teeth, and some features of modern birds, such as wings and flight feathers.

(b) Suggest how *Archaeopteryx* fossils could support the theory of evolution.

....

....

....

(2 marks)

(c) Molecular evidence for evolution has also been found.
Describe how molecular evidence can be used to support the theory of evolution by natural selection.

....

....

....

(2 marks)

3* The European corn borer is a species of moth. It is a pest that has caused large losses to corn crops in Europe and North America.

To reduce the impact of the European corn borer on corn yield, Bt corn was developed. Bt corn is a type of corn that has been genetically modified so that it produces a pesticide. The pesticide selectively kills the larvae of European corn borers when they eat it. However, natural selection may mean that some European corn borer populations have developed resistance to the pesticide produced by Bt corn.

As a result of this, farmers are advised to plant areas called 'refuge areas' next to the Bt corn crop. These refuge areas contain non-Bt corn to attract the European corn borers that have not been exposed to the Bt corn. These European corn borers can eat and lay their eggs in the refuge areas. The Bt-resistant European corn borers will also enter these refuge areas and breed with the European corn borers present there.

Using the information above, discuss how the planting of refuge areas may reduce the spread of Bt corn resistance and have implications for human populations.

(6 marks)

EXAM TIP

If you get a question like 1(b) in the exams, you must make sure your answer relates to the specific context in the question. In other words, don't just write about how natural selection works in general — you need to relate it to the evolution of fur colour in golden spiny mice in particular. Think about how having golden-brown fur might make the mice better adapted.

Score

24

Mixed Questions — 1

Here we go: a section of juicy Mixed Questions to get your teeth into. These questions draw together different bits of biology from across the entire AS-level/Year 1 course, so they should really get those brain cells firing.

For each of questions 1-3, give your answer by writing the correct letter in the box.

1 Below are four statements about the circulatory system of a fish.

Which of the statements, **A** to **D**, is **false**?

A A fish's circulatory system is closed.

B Blood passes through a fish's heart once per circuit.

C The right side of a fish's heart pumps blood to the gills, and the left side pumps blood to the rest of the body.

D A fish's veins take its blood back to the heart.

Your answer ☐

(1 mark)

2 The structural formulae of four biological molecules are displayed below.

Which of the molecules, **A** to **D**, is a pyrimidine base?

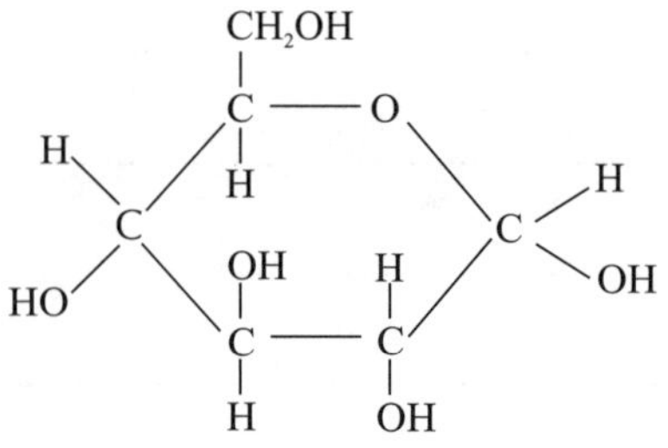

A

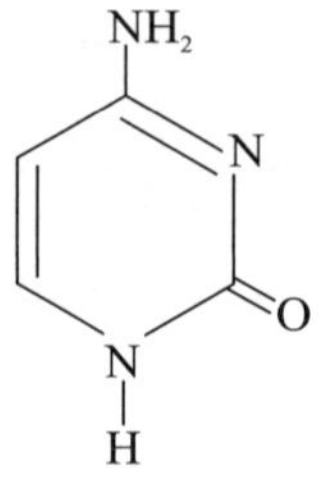

B

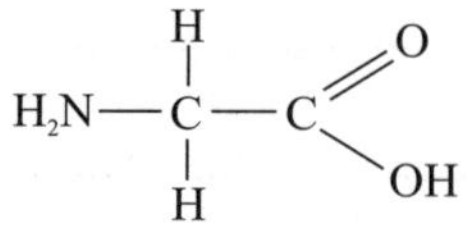

C

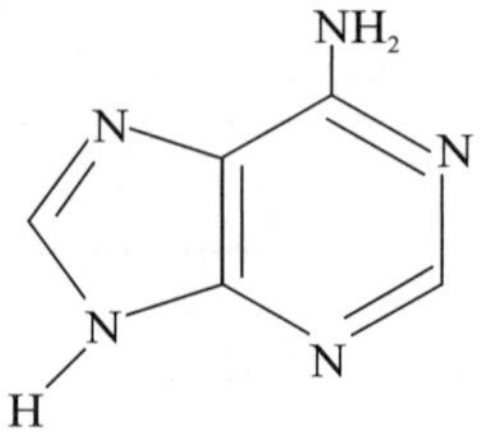

D

Your answer ☐

(1 mark)

3 The image below shows two xylem vessels.

Identify which type of microscope, **A** to **D**, was used to produce the image.

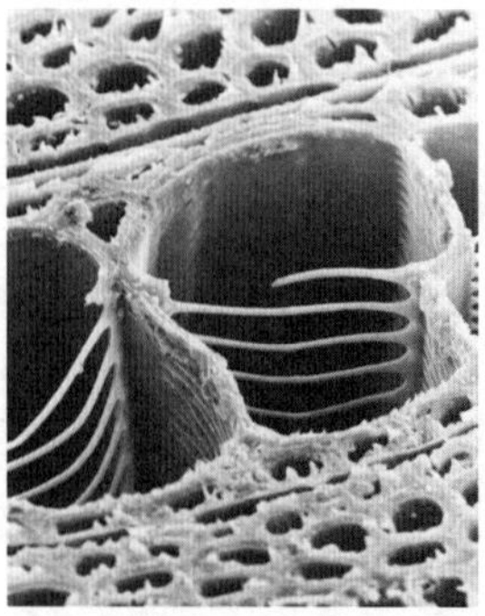

A scanning electron microscope

B light microscope

C laser scanning confocal microscope

D transmission electron microscope

Your answer ☐

(1 mark)

4 Lactase is an enzyme that catalyses the breakdown of lactose into glucose and galactose. This helps us to digest milk.

A student investigated the effect of pH on the activity of lactase, using the following method:

1. Add 2 ml of lactose solution and 2 ml of a pH buffer solution to a test tube.
   Swirl the contents of the tube gently for 10 seconds.
2. Add 2 ml of lactase solution to the test tube.
   Swirl the contents of the tube gently for 10 seconds.
3. After 8 minutes, dip a fresh glucose test strip into the solution in the test tube.
   Leave for 2 seconds and then remove.
4. After 1 minute, observe and record the colour of the test strip.
   Any shade of green indicates that glucose is present.
   If the paper remains yellow, no glucose is present.

The student repeated steps 1-4 using buffer solutions of pH 2, 4, 6, 7, 8, 10 and 12.

Table 1 shows the results of the investigation.

Table 1

pH	Glucose test result
2	Green
4	Green
6	Green
7	Green
8	Yellow
10	Yellow
12	Yellow

(a) What type of biological molecule is lactose?

....

(1 mark)

(b) (i) The solution in the test tubes contains water. Explain **one** reason why water may be needed for the reaction occurring in some of the test tubes to take place.

....

....

(2 marks)

(ii) The student's method specified that all volumes and timings should have been controlled as part of her investigation. Give **two** other variables that the student should have controlled.

1. ....

2. ....

(2 marks)

(c) (i) What can you conclude about the effect of pH on lactase activity from the results shown in **Table 1**? Explain your answer.

....

....

....

(2 marks)

(ii) Describe how enzyme structure is affected by pH.

....

....

....

....

(3 marks)

(iii) Suggest how the student could continue her investigation to determine a more accurate estimate for the pH at which lactase denatures. Explain your answer.

....

....

(2 marks)

(d) Some infants are born with a rare genetic disorder known as congenital lactase deficiency. Infants with this disorder produce either non-functioning lactase or no lactase at all. Explain how a mutation in the gene that codes for lactase could cause a non-functioning enzyme to be produced.

....

....

....

....

(3 marks)

(e) (i) Despite producing functional lactase, some people are unable to consume milk products. This is due to a milk allergy, which results in an immune response against the proteins found in milk. Describe how phagocytes could cause an immune response against milk proteins.

....

....

....

....

....

(4 marks)

(ii) Allergic reactions occur when the body's immune system reacts inappropriately to a normally harmless substance. The symptoms of an allergic reaction occur more rapidly following re-exposure to that same substance. Suggest why.

....

....

....

(2 marks)

EXAM TIP

Examiners like to put questions in contexts you've not seen before. Don't let it throw you. Instead, focus on trying to make the link between what the question's asking you and the facts you've learnt as part of your course. For example, hopefully you realised that Q4 (e) can be answered by applying what you know about immune responses to pathogens.

Score

24

Mixed Questions — 2

1 Mitosis and meiosis are both types of cell division.
Reproduction via meiosis increases genetic diversity in a population.

Figure 1 represents two cells. One cell is undergoing mitosis. The other is undergoing meiosis I.

Figure 1

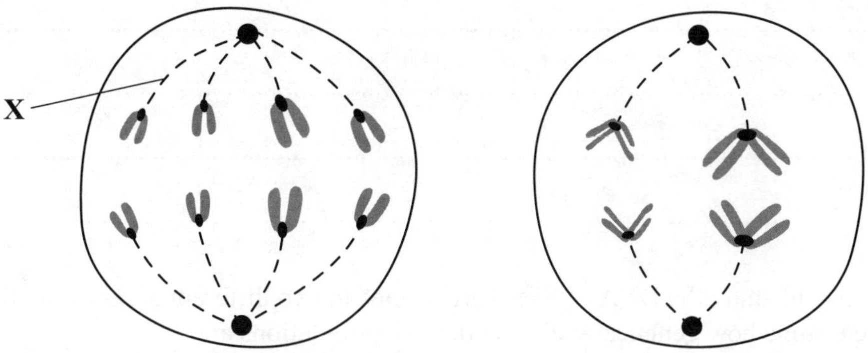

(a) (i) Identify the structure labelled **X**.

....

(1 mark)

(ii) Which of the cells in **Figure 1** (**A** or **B**) is undergoing meiosis I?
Give a reason for your answer.

....

....

(1 mark)

Figure 2 shows the life cycle of a type of plant called a liverwort. During the life cycle there are haploid (n) and diploid (2n) phases. Four stages of cell division are labelled.

Figure 2

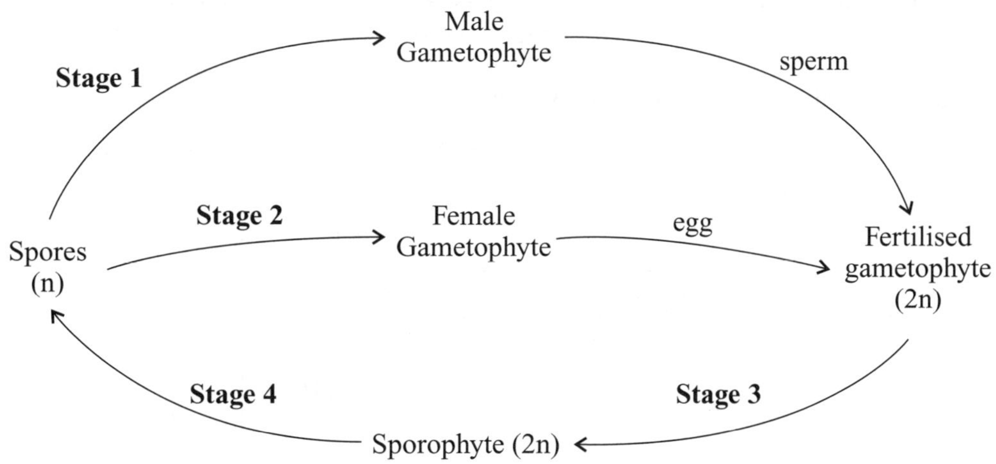

(b) The diploid number of this liverwort is 40. How many chromosomes are present in the spore?

Number of chromosomes = ...

(1 mark)

(c) (i) Name the type of cell division occurring at **Stage 4** on **Figure 2**.
Give a reason for your answer.

....

....

(1 mark)

(ii) Spores develop into male or female gametophytes (**Stages 1** and **2**).
A sperm from a male gametophyte then fertilises an egg inside a female gametophyte.
Using this information, explain why **Stages 1** and **2** on **Figure 2** must represent mitosis.

....

....

....

....

(3 marks)

(d) A scientist wants to study the DNA of liverwort species in two different areas of Scotland in order to determine how genetically diverse the two populations are.

(i) The scientist chooses to use stratified sampling to sample the liverwort populations.
Suggest and explain **one** reason why she may have chosen to use this type of sampling.

....

....

....

(2 marks)

(ii) Suggest **one** way in which the scientist could reduce the effect of random error on her results.

....

....

(1 mark)

2 *Vibrio cholerae* are pathogenic bacteria that cause the disease cholera in humans. The bacteria produce a toxin that causes Cl^- channel proteins in the ileum lining to remain open. This can result in the loss of a large volume of water from the blood.

Figure 3 shows a small section of the ileum lining.

Figure 3

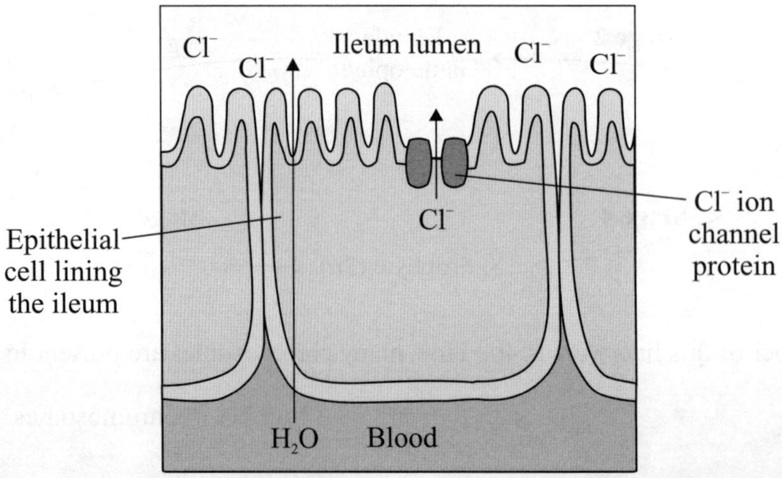

(a) Use **Figure 3** to explain how cholera causes water to be lost from the blood.

....

....

....

....

....

....

(4 marks)

(b) (i) Suggest **one** structural difference between *V. cholerae* and an ileum epithelial cell.

....

....

(1 mark)

(ii) *V. cholerae* absorbs its nutrients directly across its outer surface.
Explain why humans need a specialised exchange surface to absorb their nutrients.

....

....

....

....

....

(4 marks)

(c) Cholera outbreaks are common in areas where the treatment of sewage and drinking water is inadequate.
The disease is spread by ingesting food or water that has been contaminated with the bacteria.
The source of the contamination is usually faeces from an infected individual.
Suggest **two** ways that the spread of cholera could be reduced.

1. ....

2. ....

(2 marks)

(d) Antibiotics may be used to treat cholera. These are usually only used for severely ill patients.
Explain how this may help to ensure antibiotics could be used to treat severe cholera outbreaks in the future.

....

....

....

....

(2 marks)

EXAM TIP

Each of these Mixed Questions requires you to apply your knowledge of several different areas of the specification. It shouldn't come as a surprise that you'll get exam questions that require you to do the same thing. What it boils down to is this — you need to know your stuff inside out and back to front. So if you haven't started already, you'd better get revising.

Score

23

Answers

Module 2 : Section 1 — Cell Structure

Pages 3-5: Cell Structure — 1

1 B *[1 mark]*

The image shows a mitochondrion. The letter X is pointing to a fold in its inner membrane called a crista.

2 C *[1 mark]*

3 D *[1 mark]*

4 a) E.g. it controls the cell's activities by controlling the transcription of DNA *[1 mark]*.

b) E.g. eosin Y stains the cytoplasm so this allowed the scientist to see the cells against the background *[1 mark]*. Methylene blue and azure B stain the nucleus so this allowed the scientist to differentiate between white blood cells and red blood cells *[1 mark]*.

c) Any four from: she could have selected the lowest-powered objective lens *[1 mark]*. Then used the coarse adjustment knob to bring the stage up to just below the objective lens *[1 mark]*. Then looked down the eyepiece and used the coarse adjustment knob to move the stage down until the image is roughly in focus *[1 mark]*. Then adjusted the focus with the fine adjustment knob until she got a clear image of what's on the slide *[1 mark]*. If a greater magnification was required, she could have refocused the microscope using a higher-powered objective lens *[1 mark]*. ***[Maximum of 4 marks available]***

d) i) object size = 12 μm ÷ 1000
= 0.012 mm
magnification = image size ÷ object size
= 15 mm ÷ 0.012 mm
= **× 1250**
[accept values between × 1167 and × 1333, 2 marks for the correct answer, otherwise 1 mark for the correct working]

ii) percentage error = (uncertainty ÷ reading) × 100
uncertainty = (percentage error ÷ 100) × reading
= (0.4 ÷ 100) × 12
= **± 0.05 μm** (to 1 s.f.)
[2 marks for the correct answer, otherwise 1 mark for (0.4 ÷ 100) × 12]

5 a) Similarities: any one from: e.g. a sperm cell and a bacterial cell can both have a flagellum *[1 mark]*. / A sperm cell and a bacterial cell both have a cell membrane *[1 mark]*.
Differences: any one from: e.g. a sperm cell has a nucleus but a bacterial cell has DNA that floats freely in the cytoplasm *[1 mark]*. / A bacterial cell has a cell wall but a sperm cell only has a cell membrane *[1 mark]*.
[Maximum 2 marks available]

b) i) The flagellum requires ATP to move, which is generated by mitochondria *[1 mark]*.

ii) Electron microscope/TEM/SEM *[1 mark]* because these have a higher resolution than light microscopes, which would be needed to study the internal structures of the mitochondria *[1 mark]*.

c) E.g. microtubules are responsible for the movement of the flagellum that moves the sperm cell *[1 mark]*. The *DNAH1* mutation would result in the flagellum no longer moving/not moving properly, so the sperm cell will be unable to reach the egg and deliver genetic material *[1 mark]*.

Pages 6-8: Cell Structure — 2

1 a) E.g. each cell would only contain one nucleus. / Each nucleus would contain the same amount of genetic material. *[1 mark]*

b) The role of cell type A is to ingest invading pathogens because a greater percentage of the cell contains lysosomes than cell type B *[1 mark]*. Lysosomes are necessary to digest invading cells *[1 mark]*. The role of cell type B is to secrete enzymes because a greater percentage of the cell contains rough endoplasmic reticulum than cell type A *[1 mark]*. This organelle is covered with ribosomes which synthesise proteins, such as enzymes / is responsible for folding and processing proteins, such as enzymes *[1 mark]*.

c) E.g. chloroplasts contain thylakoid membranes/grana, whereas mitochondria contain a folded membrane that form structures called cristae *[1 mark]*. / Chloroplasts are the site of photosynthesis, whereas mitochondria are the site of aerobic respiration *[1 mark]*.

2 a) i) Put a drop of stain next to one edge of the cover slip and a piece of paper towel next to the opposite edge *[1 mark]*. The paper towel will draw the stain under the slip, across the specimen *[1 mark]*.

ii) Any one from: e.g. worn goggles/gloves / taken care not to break glass beakers/slides/cover slips / taken care with sharp tools. *[1 mark]*

b) i) One division of the stage micrometer is the same as four eyepiece divisions.
0.1 mm ÷ 4 = 0.025 mm
0.025 mm × 1000 = **25 μm**
[2 marks for the correct answer, otherwise 1 mark for the correct working]

ii) The stage micrometer will appear larger, so each eyepiece division will be a smaller measurement *[1 mark]*.

c) E.g. length of cell = 36 mm
36 mm × 1000 = 36 000 μm
object size = image size ÷ magnification
= 36 000 μm ÷ 100
= **360 μm**
[accept values between 350 μm and 370 μm, 2 marks for the correct answer, otherwise 1 mark for using the correct rearrangement of the magnification formula]

3 a) i) It is a cilium/flagellum *[1 mark]*, because the microtubules are in the 9 + 2 formation / there is an outer ring of nine microtubule pairs with two in the middle *[1 mark]*.

ii) It is involved with the separation of chromosomes during cell division *[1 mark]*.

Organelle B is a centriole.

b) i) E.g. it may cause nerve cells to lose strength/stability, causing them to break down. It may prevent vesicles from being transported through the cell to the cell surface/plasma membrane to secrete their contents. ***[1 mark each]***

ii) E.g. abnormal function of mitochondria in nerve cells may result in the cell not receiving enough energy, causing them to die/preventing them from carrying out their normal function *[1 mark]*.

Module 2 : Section 2 — Biological Molecules

Pages 9-12: Biological Molecules — 1

1 D ***[1 mark]***

The formula for a phosphate ion is PO_4^{3-}.
It has a negative charge, so it's an anion.

2 C ***[1 mark]***

The protein backbone will have peptide bonds between the amino acids and the carbohydrate branches will have glycosidic bonds between the monosaccharides.

3 A ***[1 mark]***

4 a) A molecule made from many monomers bonded together ***[1 mark]***.

b) i) E.g.

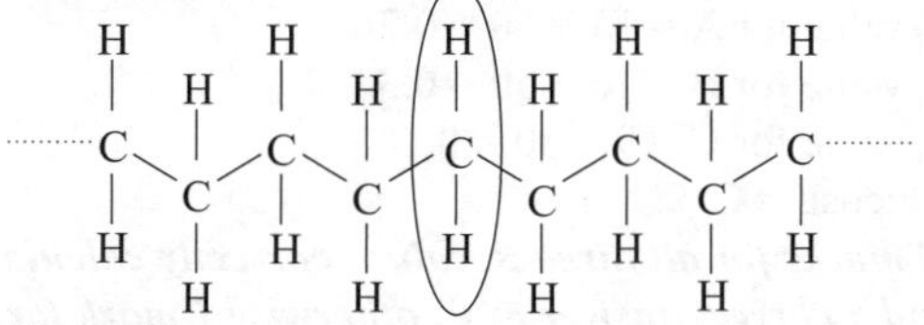

[1 mark]

ii) E.g. monosaccharides and amino acids ***[1 mark]***.

5 a) i) alpha-glucose/α-glucose ***[1 mark]***

You must have specified alpha-glucose to get the mark here. Make sure you know the difference between alpha- and beta-glucose.

ii) maltose ***[1 mark]***

iii) Number of carbon atoms = 12 ***[1 mark]***

Each alpha-glucose contains 6 carbon atoms (it's a hexose sugar).

b) A water molecule is used ***[1 mark]*** to break/hydrolyse the glycosidic bond in the disaccharide/maltose ***[1 mark]***. This produces the two monomers/alpha-glucose molecules ***[1 mark]***.

6 a) E.g. add a few drops of sodium hydroxide to a solution of the sample, then a few drops of copper (II) sulfate solution / add a few drops of biuret reagent to the sample ***[1 mark]***. A colour change from blue to purple/lilac is a positive result ***[1 mark]***.

b) i) E.g. haemoglobin ***[1 mark]***

ii) So that it can be transported in the blood to the tissues where it acts ***[1 mark]***.

Insulin is a hormone. Hormones are secreted by glands and transported to their target cells by the blood, so they must be soluble.

iii) Hydrophobic R groups on the amino acids of the insulin molecule clump together ***[1 mark]***. This causes hydrophilic R groups to be pushed to the outside of the molecule, making the molecule soluble ***[1 mark]***.

iv) E.g. insolubility ***[1 mark]*** and strength ***[1 mark]***.

7 a) A lot of energy is removed from the kangaroo's body when the water in the saliva evaporates from its forearms ***[1 mark]***. This reduces the kangaroo's body temperature ***[1 mark]***.

b) i) Water has a high specific heat capacity / it takes a lot of energy to break the hydrogen bonds in water ***[1 mark]***, which means it doesn't heat up as quickly as the air ***[1 mark]***.

ii) The water molecules are polar ***[1 mark]***, which means that hydrogen bonds form between them ***[1 mark]***. This means the water molecules are cohesive, which allows them to travel up a tree trunk ***[1 mark]***.

8 a) i)

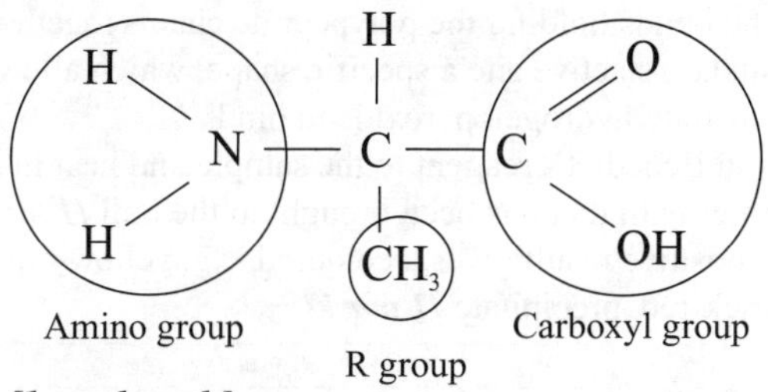

[1 mark each]

ii) It has a different R group ***[1 mark]***.

b) i) water ***[1 mark]***

ii) E.g.

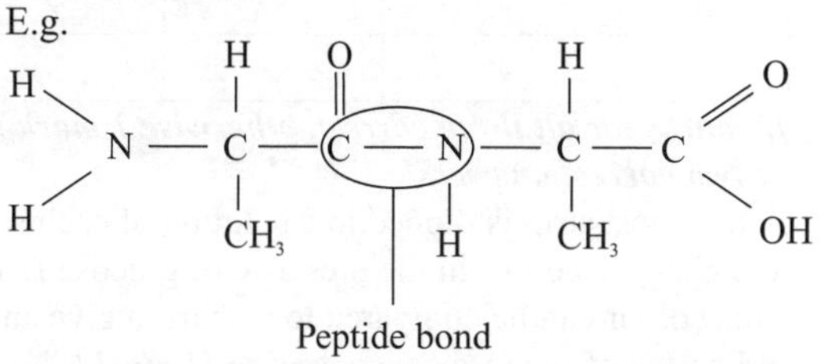

[1 mark for the correct diagram, 1 mark for correctly labelling the peptide bond]

Pages 13-16: Biological Molecules — 2

1 a) i) alpha-helix/α-helix ***[1 mark]***

ii) Both secondary and tertiary structures contain hydrogen bonds ***[1 mark]***. Tertiary structures also contain ionic bonds / disulfide bridges / hydrophobic and hydrophilic interactions ***[1 mark]***.

b)* How to grade your answer:

Level 0: There is no relevant information. ***[No marks]***

Level 1: The answer briefly explains the structure of catalase. The answer has no clear structure. The information given is basic and lacking in detail. It may not all be relevant. ***[1 to 2 marks]***

Level 2: The answer explains the structure of catalase and makes at least one suggestion about how the structure of catalase helps it to carry out its function. The answer has some structure. Most of the information given is relevant and there is some detail involved. ***[3 to 4 marks]***

Level 3: The answer fully explains the structure of catalase, given the information provided, and makes several suggestions about how this helps the enzyme to carry out its function. The answer has a clear and logical structure. The information given is relevant and detailed. ***[5 to 6 marks]***

Indicative scientific content may include:

<u>The structure of catalase</u>

Catalase is a globular protein, so it is round and compact.
Its hydrophilic R groups are pushed to the outside of the protein.
Catalase is conjugated, so it contains (a) non-protein/prosthetic group(s).
Catalase has a quaternary structure, so it is made up of more than one polypeptide chain.
The polypeptide chains are held together by bonds.

<u>How the structure helps catalase carry out its function</u>

Its compact size allows catalase to act in the peroxisome, which is relatively small.
The hydrophilic R groups on the outside of the protein make the enzyme soluble, so it can be easily transported from where it is made in the cytoplasm to where it acts.
The non-protein / prosthetic group(s) may help catalase to bind to hydrogen peroxide.

The bonds holding the polypeptide chains together give catalase's active site a specific shape, which allows the substrate/hydrogen peroxide to bind.

2 a) i) Add Benedict's reagent to the sample and heat in a water bath that has been brought to the boil ***[1 mark]***. A positive result gives a coloured, e.g. yellow/orange/brick-red, precipitate ***[1 mark]***.

ii)

Sample	Type of carbohydrate present		
	Reducing sugar	Non-reducing sugar	Starch
A		✓	
B			✓
C	✓		

[2 marks for all three correct, otherwise 1 mark for one or two correct answers]

b) i) A fresh test strip is dipped in a solution of each sample and changes colour in the presence of glucose ***[1 mark]***. This colour can be compared to a chart to give an indication of glucose concentration ***[1 mark]***.

ii) E.g. colorimetry / a glucose biosensor ***[1 mark]***

3 a) phospholipid ***[1 mark]***

b) i) The phospholipid head is hydrophilic ***[1 mark]***, but the phospholipid tail is hydrophobic ***[1 mark]***. The molecules arrange themselves in this way to shield the hydrophobic tails from water ***[1 mark]***.

ii) E.g. they make up cell membranes ***[1 mark]***. The hydrophobic regions act as a barrier to water-soluble substances ***[1 mark]***.

4 a) i) Fatty acid 2 has a double bond between two carbon atoms, whereas fatty acid 1 does not ***[1 mark]***. This means that fatty acid 1 is saturated and fatty acid 2 is unsaturated ***[1 mark]***.

ii) Three fatty acids combine with one glycerol molecule ***[1 mark]*** in a series of three condensation reactions / a process known as esterification ***[1 mark]***. Ester bonds are formed between the glycerol and fatty acids ***[1 mark]***.

iii) E.g. triglycerides are used for storage of energy ***[1 mark]*** because they have a high energy content / they're insoluble ***[1 mark]***.

b) i) Shake some of the sample to be tested with ethanol until it dissolves, then pour into a test tube of water ***[1 mark]***. If a lipid is present, a white emulsion will form ***[1 mark]***.

ii) Lipids cannot dissolve in water so if they are present, they form an emulsion ***[1 mark]***.

Pages 17-19: Biological Molecules — 3

1 a) i) Alpha-glucose molecules join together in a chain via a serious of condensation reactions ***[1 mark]***, which form glycosidic bonds between the molecules ***[1 mark]***.

ii) E.g.

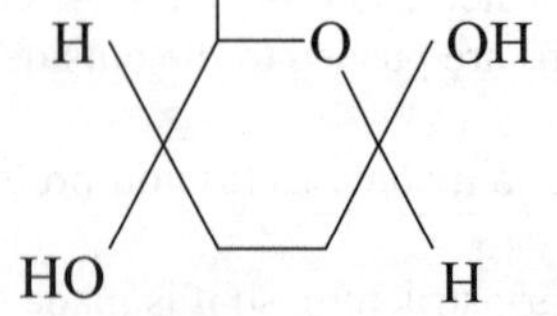

[1 mark]

In alpha-glucose the OH/hydroxyl group and the H/hydrogen on the right-hand side are reversed ***[1 mark]***.

You could have drawn the full skeletal formula for beta-glucose here instead.

iii) It allows cellulose molecules to form strong fibres/microfibrils ***[1 mark]***, which provide structural support/support the cell wall ***[1 mark]***.

b) Advantage: e.g. it is more compact, which allows more of it to be stored in a cell ***[1 mark]***.
Disadvantage: e.g. it can't be broken down as quickly, so the glucose it stores is less readily available ***[1 mark]***.

2 a) i) To stop the solvent evaporating ***[1 mark]***.

ii) To act as the mobile phase, through which the molecules that are being separated can move ***[1 mark]***.

iii) Water is also a polar molecule ***[1 mark]***. This means that the glucose molecules are attracted to the water molecules / can form hydrogen bonds with the water molecules and will dissolve in water ***[1 mark]***.

b) i) $R_f = \frac{\text{distance travelled by solute}}{\text{distance travelled by solvent}}$

Distance travelled by solvent = 50 mm
R_f value for A = 13 ÷ 50 = 0.26
R_f value for B = 16 ÷ 50 = 0.32
R_f value for C = 9 ÷ 50 = 0.18
Glucose = **C**

[2 marks for all three R_f values correctly calculated and a correct answer of C, otherwise 1 mark for two R_f values calculated correctly]

ii) To make the spots of colourless sugars visible ***[1 mark]***.

3 a)

Final concentration of glucose /mM	Volume of distilled water / cm^3	Volume of glucose solution / cm^3
4.0	0.0	10.0
2.0	5.0	5.0
1.0	7.5	2.5
0.5	8.75	1.25

[1 mark for three correct glucose concentrations]

b) Any four from: the student could carry out the Benedict's test on each known glucose solution ***[1 mark]***. He could then use the colorimeter to measure the absorbance values of the solutions resulting from each Benedict's test ***[1 mark]***. These absorbance values could then be used to plot a calibration curve ***[1 mark]***. A Benedict's test could then be carried out on a sample of each energy drink and a colorimeter measurement taken ***[1 mark]***. The absorbance value for each energy drink sample would then be read off the calibration curve to find the associated glucose concentration ***[1 mark]***. ***[Maximum of 4 marks available]***

c) i) E.g. 10 cm^3/the same volume of distilled water only ***[1 mark]***.

ii) To demonstrate that only the colour of the solution/the presence of glucose is affecting the readings ***[1 mark]***.

d) E.g. the colour of the drinks may have increased the absorbance readings ***[1 mark]***, and therefore the estimates of glucose concentration for these drinks will be too low ***[1 mark]***.

Module 2 : Section 3 — Nucleotides and Nucleic Acids

Pages 20-22: Nucleotides and Nucleic acids — 1

1 C ***[1 mark]***
2 D ***[1 mark]***
3 D ***[1 mark]***

*First work out what the DNA mutation is: TAC/**CTA**/GCT has become TAC/**CTC**/GCT during replication. Now work out the change to the mRNA sequence: the complementary mRNA codon for CTA is GAU, and the complementary mRNA codon for CTC is GAG. Then you can use the table to find that GAU codes for Asp (aspartate), and GAG codes for Glu (glutamate).*

4 a) i) Ribose ***[1 mark]***
ii) Phosphodiester bond ***[1 mark]***
b) i) The molecule contains uracil/U bases (in place of thymine/T bases) ***[1 mark]***.
ii) Complementary base pairing ***[1 mark]*** means that hydrogen bonds will form between the base pairs A and U, and C and G ***[1 mark]***. Because the two halves of the RNA sequence are complementary, it causes the RNA strand to fold into a stem-loop structure ***[1 mark]***.
iii) Number of purines (A and G) = 12
Total bases = 23
$(12 \div 23) \times 100 =$ **52.2%** ***[1 mark]***

5 a) i) DNA helicase separates the nucleotide strands / causes the DNA helix to unwind ***[1 mark]*** by breaking the hydrogen bonds between bases ***[1 mark]***. DNA polymerase joins the nucleotides in the new DNA strand together ***[1 mark]***.
ii) To ensure genetic information is conserved/stays the same/is not changed ***[1 mark]***. Changes to the DNA sequence could alter the sequence of amino acids in a protein ***[1 mark]***, which may result in the protein no longer functioning properly ***[1 mark]***.
b) i) The results show that DNA replication has not occurred in the absence of ATP or when ATP hydrolase is inactive ***[1 mark]***.
ii) E.g. the results suggest that the breakdown of ATP (by ATP hydrolase) is essential for DNA replication ***[1 mark]***. This could be because DNA replication requires energy and/or inorganic phosphate released by the breakdown of ATP ***[1 mark]***.

Pages 23-25: Nucleotides and Nucleic acids — 2

1 a) E.g. valine is coded for by four DNA codons ***[1 mark]***.
'Degenerate' means that multiple DNA codons can code for one amino acid, so any evidence that shows this from the table could gain a mark.
b) i) An A will be swapped for a C / the mRNA codon GAC will become GCC ***[1 mark]***.
ii) The fifth amino acid in this part of the sequence will be arginine instead of a leucine ***[1 mark]***.
c) The mRNA attaches to a ribosome. tRNA molecules carry **amino acids** ***[1 mark]*** to the ribosome. tRNA molecules attach to the mRNA via **complementary base pairing** ***[1 mark]***. rRNA in the ribosome catalyses the formation of a **peptide bond** ***[1 mark]*** between two **amino acids** ***[1 mark]***. The ribosome moves along the mRNA, producing a polypeptide chain. The process continues until a **stop codon** ***[1 mark]*** on the mRNA is reached.

2 a) Any four from: e.g. after the DNA has been unwound ***[1 mark]***, RNA polymerase lines up free RNA nucleotides along the template strand ***[1 mark]*** according to the rules of complementary base pairing ***[1 mark]***. RNA polymerase then joins the RNA nucleotides together as it moves along the DNA template strand ***[1 mark]***. Once RNA polymerase reaches a stop codon, it detaches and the mRNA is released ***[1 mark]***. ***[Maximum of 4 marks available]***
b) i) The levels of mRNAs 2 and 3 were only slightly affected by the drug ***[1 mark]***. If RNA polymerase was inhibited, the levels of all of the mRNAs would have decreased by a large amount ***[1 mark]***.
ii) Only the level of mRNA 1 has been significantly reduced ***[1 mark]***. This indicates that mRNA 1 contains the particular sequence destroyed by the drug ***[1 mark]***.

3 How to grade your answer:
Level 0: There is no relevant information. ***[No marks]***
Level 1: There is a brief description of semi-conservative replication or an attempt to explain how the experiment provides evidence that DNA replication is a semi-conservative process. The answer has no clear structure. The information given may be basic and lacking in detail. It may not all be relevant. ***[1 to 2 marks]***
Level 2: There is a description of semi-conservative replication and some explanation of how the experiment provides evidence that DNA replication is a semi-conservative process. The answer has some structure. Most of the information given is relevant and there is some detail involved. ***[3 to 4 marks]***
Level 3: There is a description of semi-conservative replication and a full explanation of how the experiment provides evidence that DNA replication is a semi-conservative process. The answer has a clear and logical structure. The information given is relevant and detailed. ***[5 to 6 marks]***

Indicative scientific content may include:
<u>Knowledge of semi-conservative replication</u>
Original DNA molecule unwinds.
Each strand acts as a template for a new DNA strand, which is made from free-floating DNA nucleotides.
Each new DNA molecule contains one original (template) strand and one new strand.
<u>Explanation of results as evidence for semi-conservative replication</u>
100% of the DNA molecules in bacterial generation 0 contained heavy nitrogen. This is because the *E. coli* were originally grown in a heavy nitrogen nutrient broth.
100% of the DNA molecules in bacterial generation 1 contained both heavy nitrogen and light nitrogen.
This suggests that DNA replication could be semi-conservative, with each DNA molecule in generation 1 containing one heavy nitrogen template strand (from bacterial generation 0) and one new strand made up of light nitrogen nucleotides that were synthesised from the surrounding broth.
In bacterial generation 2, 50% of the DNA contained both heavy and light nitrogen and 50% contained only light nitrogen.

This means that DNA replication must be semi-conservative because any DNA molecules synthesised from a template strand containing heavy nitrogen will contain both heavy and light nitrogen.
Also, any DNA molecules synthesised from a template strand that only contains light nitrogen will themselves only contain light nitrogen.
The percentage of DNA containing both heavy and light nitrogen halves with each generation, confirming that DNA replication must be semi-conservative.

Module 2 : Section 4 — Enzymes

Pages 26-27: Enzymes — 1

1 C ***[1 mark]***

At temperatures below the optimum, a Q_{10} value of 2 means the reaction rate doubles when the temperature is raised by 10 °C. 0.8 × 2 = 1.6.

2 B ***[1 mark]***

Competitive inhibitors compete with the substrate for an enzyme's active site. This means the rate of reaction is slower at lower substrate concentrations than it would have been without the inhibitor. The maximum rate of reaction is still reached, but it takes a greater concentration of substrate than is needed without the inhibitor.

3 a) At low temperatures, the rate of the reaction is slow because the kinetic energy of the enzyme and substrate molecules is low ***[1 mark]***. As temperature increases to the optimum, the rate increases as there are more successful collisions between enzymes and substrate molecules ***[1 mark]***. At temperatures higher than the optimum, the rate decreases as the enzyme is denatured ***[1 mark]***.

b) Enzyme B because it has the higher optimum temperature ***[1 mark]***, which will allow it to function at the higher temperatures found in a tropical climate ***[1 mark]***.

c) The insecticide molecule is a similar shape to enzyme A's substrate ***[1 mark]***. The insecticide molecule occupies enzyme A's active site ***[1 mark]*** so the substrate cannot fit and the respiration reaction cannot be catalysed ***[1 mark]***. This interrupts the respiration process and kills the insect ***[1 mark]***.

Pages 28-30: Enzymes — 2

1 a) Pepsin is found in the body, so 40 °C is likely to be within the optimal range for pepsin action ***[1 mark]***.

b) The reaction is fastest at pH 2 ***[1 mark]***, which suggests pepsin has an optimum pH between those values found in the stomach (pH 1.5 to pH 3.5) ***[1 mark]***. At pH values similar to those found in the small intestine (pH 5 and pH 8), the reaction is much slower ***[1 mark]***, suggesting that pepsin would denature in the small intestine ***[1 mark]***.

2 a) i) Extracellular enzymes work outside cells, whereas intracellular enzymes work within cells ***[1 mark]***.

ii) E.g. amylase / trypsin ***[1 mark]***.

b) The independent variable is enzyme concentration and the dependent variable is the volume of oxygen produced ***[1 mark]***.

c) ± 0.5 cm^3 ***[1 mark]***

d) E.g. collected the gas in a measuring cylinder with a smaller resolution ***[1 mark]***.

e) i) Rate = $\frac{17}{20}$

= **0.85 $cm^3\ s^{-1}$**

[1 mark for 0.85, 1 mark for the correct units]

ii)

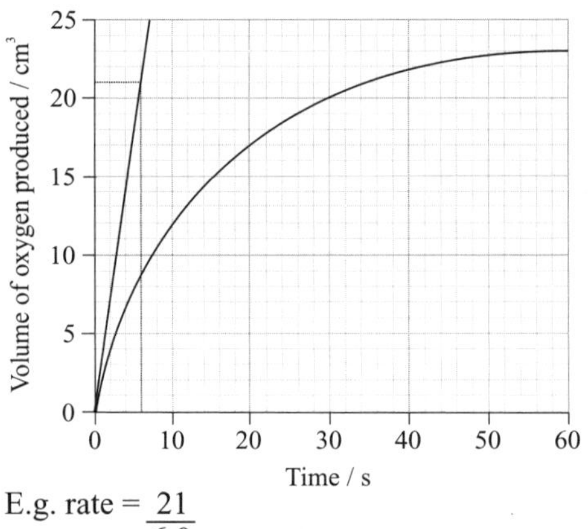

E.g. rate = $\frac{21}{6.0}$

= **3.5 $cm^3\ s^{-1}$**

[1 mark for a tangent drawn to the curve at time = 0 s, 1 mark for an answer between 3.2 $cm^3\ s^{-1}$ and 3.8 $cm^3\ s^{-1}$]

Tangents can be tricky things to draw accurately so the examiners will usually accept answers that are a bit below or a bit above what they got themselves — even so, try to draw your line as carefully as you can.

f) E.g.

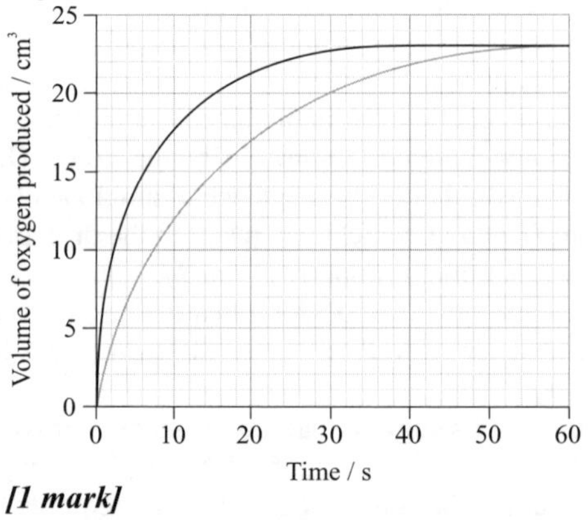

[1 mark]

Don't worry if your line's not exactly like this. You just need to make sure it's steeper than the original but still starts and plateaus at the same values.

3 a) Vitamin KO binds to the active site of vitamin KO reductase ***[1 mark]***. As it does so, vitamin KO makes the active site change shape slightly to fit more closely around it ***[1 mark]***. Reduced vitamin K is not able to make vitamin KO reductase's active site change shape in the right way, so vitamin KO reductase won't catalyse its conversion back to vitamin KO ***[1 mark]***.

b) It is an organic, non-protein molecule ***[1 mark]***, that binds to an enzyme and helps it to function ***[1 mark]***.

c) By inhibiting the action of vitamin KO reductase, warfarin prevents the production of the coenzyme reduced vitamin K ***[1 mark]***. Without reduced vitamin K, carboxylase is unable function properly and can't catalyse the conversion of precursor prothrombin into prothrombin ***[1 mark]***. This means that thrombin and fibrin can't be made ***[1 mark]*** and the lack of fibrin prevents a blood clot from forming ***[1 mark]***.

Module 2 : Section 5 — Biological Membranes

Pages 31-34: Biological Membranes — 1

1 C ***[1 mark]***

2 B ***[1 mark]***

Hypotonic means the solution has a higher water potential than the cells. Water moves from areas of higher water potential to areas of lower water potential, so water will move from the solution into the cells. The cells swell and become turgid.

3 D ***[1 mark]***

Ammonia will diffuse fastest when the surface area to volume ratio of the cube is highest and its concentration gradient is steepest. Smaller cubes have a higher surface area to volume ratio and the more concentrated ammonia solution will provide the steepest concentration gradient (i.e. the biggest difference in ammonia concentration).

4 a) i) Proteins are scattered amongst the phospholipids, like tiles in a mosaic ***[1 mark]***. The phospholipids are constantly moving, so the structure is fluid ***[1 mark]***.

ii) The cholesterol molecules would restrict the movement of the phospholipids ***[1 mark]***, making the structure less fluid and more rigid ***[1 mark]***.

iii) Any two from: e.g. they stabilise the membrane by forming hydrogen bonds with surrounding water molecules. / They bind to drugs/hormones/antibodies. / They act as receptors for cell signalling. / They act as antigens. ***[2 marks]***

iv) Any two from: e.g. phospholipids would be packed more closely together. / The channel proteins and carrier proteins would deform. / Ice crystals may form and pierce the membrane. ***[2 marks]***.

b) i) E.g. a large number of carrier or channel proteins ***[1 mark]*** in order to allow cations to cross the cell membrane quickly ***[1 mark]***.

ii) E.g. at lower temperatures, sodium ions have less kinetic energy ***[1 mark]***, so diffuse across the nerve cell membrane at a slower rate, which slows the conduction of the nerve impulse ***[1 mark]***.

5 a) B ***[1 mark]***

b) The hydrophobic part of the bilayer prevents water-soluble/charged particles from diffusing through it ***[1 mark]***. This allows the membrane to control what enters and leaves the cell ***[1 mark]***.

c) The water will move from the exterior to the interior of the cell ***[1 mark]*** because the water potential of the exterior is higher/less negative than the water potential of the interior ***[1 mark]***.

6 a) Liver and skeletal muscle cells have different types of receptors on their surface, each with a specific shape ***[1 mark]***. Glucagon has a complementary shape to a liver cell receptor but not a skeletal muscle cell receptor, so it is able to bind to liver cells, and not to skeletal muscle cells ***[1 mark]***.

b) It is a similar shape to glucagon, so it may bind to glucagon-specific receptors in the place of glucagon ***[1 mark]***. This will prevent glucagon from activating the conversion of glycogen into glucose and increasing the blood sugar level ***[1 mark]***.

7 a) To make sure any betalains/pigments released by the cutting of the beetroot were washed away ***[1 mark]***.

b) Colorimetry analysis of distilled water ***[1 mark]***.

c) Any four from: e.g. increasing the temperature from 20 °C to 40 °C increases the fluidity of the phospholipids in the beetroot cell membranes ***[1 mark]***. At temperatures above 40 °C, the membrane starts to break down / proteins in the membrane start to denature ***[1 mark]***. The membrane surrounding the vacuole therefore becomes more permeable with increasing temperature ***[1 mark]***, meaning that betalains/pigments leak out into the distilled water ***[1 mark]***. The more pigments released, the higher the absorbance reading ***[1 mark]***.

d) Cell membranes contain channel proteins and carrier proteins ***[1 mark]***. Proteins are denatured by extremes of pH / extremes of pH interfere with the bonding in proteins, causing them to change shape ***[1 mark]***. If the proteins are not able to function and control what goes in or out of the cell, membrane permeability will increase ***[1 mark]***.

Pages 35-36: Biological Membranes — 2

1 a) E.g. histamine is too big / it's charged/polar ***[1 mark]***.

You're not expected to know the characteristics of histamine, just to apply your knowledge of why molecules are usually unable to diffuse directly through cell membranes.

b) Exocytosis ***[1 mark]***. The granules containing histamine fuse with the plasma membrane and release their contents outside of the cell ***[1 mark]***.

2 a)

Concentration of sucrose solution to be made up / mol dm^{-3}	Volume of 1 mol dm^{-3} sucrose solution used / cm^3	Volume of water used / cm^3	Final volume of solution to be made up / cm^3
1	20	0	20
0.75	15	**5**	20
0.5	**10**	**10**	20
0.25	**5**	**15**	20
0	**0**	**20**	20

[2 marks for all four rows correct, otherwise 1 mark for three rows correct]

b) Any two from: e.g. the temperature the potato samples were incubated at / the length of time the potato samples were incubated for / the volume of sucrose solution used / the variety of potato used / the age of potato used. ***[2 marks]***

c) The line of best fit crosses the x-axis of Figure 1 halfway between 0.25 and 0.50, so the sucrose concentration of potato cells = approximately 0.375 mol dm^{-3}.
A 0.3 mol dm^{-3} sucrose solution has a water potential of –850 kPa. A 0.4 mol dm^{-3} sucrose solution has a water potential of –1130 kPa.
So a 0.375 mol dm^{-3} sucrose solution has a water potential of approximately:
$(-1130) - (-850) = 280 \times 0.75 = 210$
$-850 - 210 =$ **–1060 kPa**
[2 marks for an answer > –850 and < –1130 kPa, otherwise 1 mark for estimating the sucrose concentration of the potato cells to be between 0.3 and 0.4 mol dm^{-3}]

d) The sweet potato tissue is likely to have a lower water potential than that of the white potato ***[1 mark]*** because it is likely to have a higher sucrose concentration ***[1 mark]***.

The extra sucrose (with some other sugars too) is what makes the sweet potato sweet.

e) No, the cells in pure water increased in mass, so they gained water ***[1 mark]***. Cells become plasmolysed when they lose water, not when they gain it ***[1 mark]***.

Module 2 : Section 6 — Cell Division and Cellular Organisation

Pages 37-40: Cell Division and Cellular Organisation

1 B ***[1 mark]***

2 C ***[1 mark]***

3 A ***[1 mark]***

Chromosomes attach to the spindle during metaphase of mitosis (M phase) and remain attached until telophase. So the attachment of the chromosomes to the spindle must be checked during M phase.

4 a) Root tips are actively growing/contain meristems, so the cells here will be undergoing mitosis/dividing ***[1 mark]***.

b) It means that the cells have been deliberately squashed beneath the cover slip ***[1 mark]***. It was used because it makes it easier to see the chromosomes ***[1 mark]***.

c) i) E.g. if a cell contains visible/condensed chromosomes this indicates that it is dividing ***[1 mark]***.

ii) 80 + 240 = 320 cells in total
percentage undergoing mitosis =
(240 ÷ 320) × 100 = **75%** ***[1 mark]***

d) i) Any two from: e.g. stem cells are unspecialised cells. / Stem cells are capable of differentiating to form other types of cell. / Stem cells are capable of self-renewing. ***[2 marks]***

ii) E.g. xylem / phloem ***[1 mark]***

5 a) i) S phase ***[1 mark]***. Blocking the formation of nucleotides will stop DNA from replicating ***[1 mark]***.

ii) The cells will be unable to pass through the checkpoint that checks DNA has been successfully replicated/the next checkpoint/the G_2 checkpoint ***[1 mark]***.

b) E.g. red blood cells are produced from stem cells in the bone marrow ***[1 mark]***. Methotrexate stops the cell cycle, so it may prevent stem cells from dividing to produce more red blood cells (leading to a low red blood cell count) ***[1 mark]***.

6 a) i) In anaphase II, the sister chromatids are separated ***[1 mark]***, whereas in anaphase I the pairs of homologous chromosomes are separated ***[1 mark]***.

ii) Telophase II occurs / the nuclear envelope forms around the separate sets of chromatids ***[1 mark]***. Cytokinesis then occurs / the cytoplasm divides (to produce four daughter cells) ***[1 mark]***.

b) A pollen grain is a gamete and therefore a haploid cell ***[1 mark]***. Meiosis is needed to halve the chromosome number and produce a haploid cell from a diploid cell ***[1 mark]***.

c) The crossing over of homologous chromosomes in meiosis I ***[1 mark]*** leads to the four daughter cells containing different combinations of alleles ***[1 mark]***. Independent assortment in meiosis I ***[1 mark]*** leads to the daughter cells containing any combination of maternal and paternal chromosomes ***[1 mark]***.

7 a) Prophase ***[1 mark]***. The chromosomes have condensed but have not started to divide ***[1 mark]***.

b) A peak in the concentration of cyclin E occurs when the mass of DNA starts to increase ***[1 mark]***. This suggests that cyclin E may trigger DNA replication in the cell / entry into the S stage of interphase ***[1 mark]***. The peak in the concentration of cyclin B is followed by a decrease/halving in the mass of DNA ***[1 mark]***. This suggests that cyclin B may trigger the cell to enter mitosis/M phase ***[1 mark]***.

Module 3 : Section 1 — Exchange and Transport

Pages 41-43: Exchange and Transport — 1

1 B ***[1 mark]***

2 C ***[1 mark]***

Tidal volume is the volume of air in an ordinary breath in or out — so in this graph, it's the height of one of the small peaks.

3 A ***[1 mark]***

4 a) E.g. dissecting scissors ***[1 mark]***

b) i) A: spiracle ***[1 mark]***
B: tracheae ***[1 mark]***

ii) E.g. pipette a drop of water onto a slide ***[1 mark]***. Use tweezers to place a section of structures B/the tracheae onto the drop of water ***[1 mark]***. Stand a cover slip upright on the slide, next to the water drop, then carefully tilt and lower it so it covers the specimen ***[1 mark]***.

c) It changes the volume of its body with rhythmic abdominal movements ***[1 mark]***. This changes the pressure in its body so that air is drawn in and out through the spiracles ***[1 mark]***.

d) E.g. oxygen from the air dissolves in the (tracheal) fluid, then diffuses from the fluid into the body cells ***[1 mark]***. Carbon dioxide diffuses into the fluid in the opposite direction ***[1 mark]***.

5 a) i) gill plate/(secondary) lamella ***[1 mark]***

ii) An arrow drawn across structure X in the opposite direction to the arrow showing water flow across the gill filament, e.g.

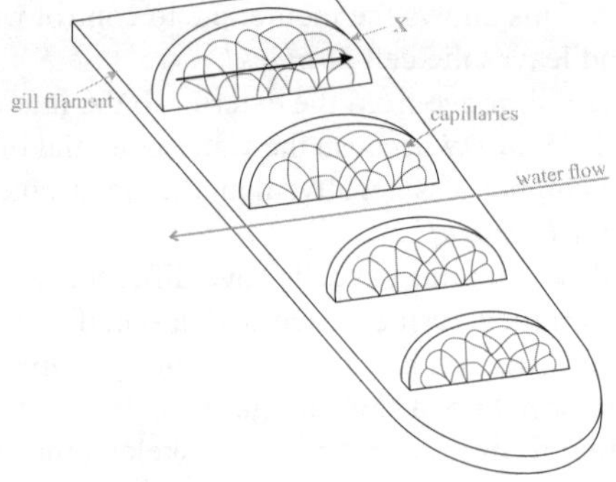

[1 mark]

Fish gills have a counter-current system, meaning the blood flows in the opposite direction to the water.

iii) E.g.

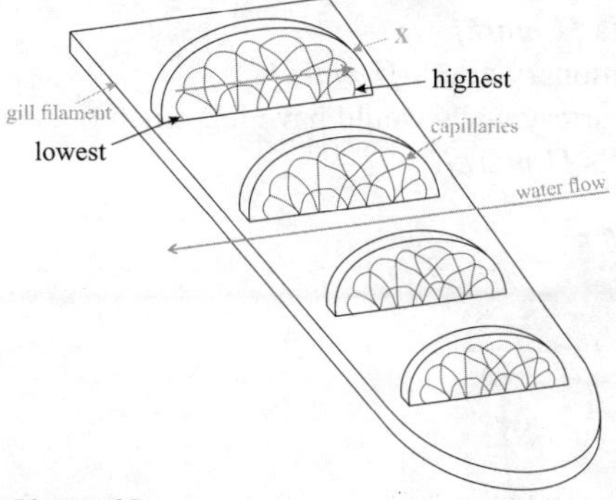

[1 mark]

Blood enters the gill plate with a low oxygen concentration. Because a steep concentration gradient is maintained between the water and the blood, a lot of oxygen has diffused into the blood by the time it leaves the gill plate.

b) E.g. the many gill plates/(secondary) lamellae give the gill filament a large surface area ***[1 mark]***, increasing the rate of diffusion of gases ***[1 mark]***.

c) E.g. when the fish opens its mouth, the floor of the buccal cavity is lowered, increasing the cavity's volume ***[1 mark]***. This decreases the pressure in the buccal cavity, so water is drawn into it ***[1 mark]***. When the fish closes its mouth, the floor of the buccal cavity is raised, decreasing the volume of the buccal cavity ***[1 mark]***. This increases the pressure in the cavity, so water is forced out across the gill filaments ***[1 mark]***.

Pages 44-46: Exchange and Transport — 2

1 a) E.g.
Bacterium A: 5 : 1
Bacterium B: 84 ÷ 22 = 3.8181..., so 84 : 22 = 3.81 : 1
Bacterium C: 15.75 ÷ 3.5 = 4.5, so 15.75 : 3.5 = 4.5 : 1
Bacterium with fastest gas exchange: Bacterium **A**
[1 mark for bacterium A, 1 mark for 3.81 or 4.5]

Converting all the ratios into the form n : 1 makes it easier to see that Bacterium A has a bigger surface area compared to its volume, which means there is a larger surface for gases to diffuse across. This means the rate of gas exchange can be faster.

b) Any two from: e.g. it would take too long for substances to travel from the outer surface to the cells deep within the body ***[1 mark]***. / Not enough substances would be exchanged across the outer surface, which is small relative to the volume of a multicellular organism ***[1 mark]***. / Multicellular organisms have a higher metabolic rate than single-celled organisms, so diffusion across the outer surface would be too slow to keep up with metabolic demand ***[1 mark]***.

2 a)

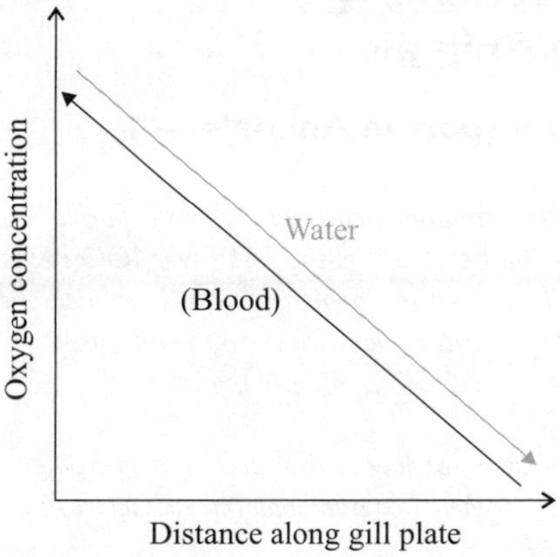

[1 mark]

b) In the parallel flow system, the oxygen concentration gradient between the water and the blood decreases with distance along the gill plate ***[1 mark]***, which will mean that the rate of oxygen diffusion will also decrease with distance along the gill plate ***[1 mark]***. This means that less oxygen diffuses into the blood than with a counter-current system, where the concentration gradient (and therefore rate of diffusion) is maintained ***[1 mark]***.

3 a) 3 seconds ***[1 mark]***

Lung volume will be smallest when the pressure is highest (and the person is expiring).

b) 0-2 seconds ***[1 mark]*** because this is when the pressure inside the lungs/the intrapulmonary pressure is negative ***[1 mark]***. Air travels from an area of higher pressure to an area of lower pressure, so air must be being taken into the lungs at this point ***[1 mark]***.

c) 60 ÷ 4 = 15
15 breaths min^{-1} ***[1 mark]***

The graph shows one full breath (inspiration and expiration) which takes 4 seconds to complete.

4 a) Any three from: e.g. less oxygen is able to reach the alveoli because the bronchi are swollen/filled with mucus ***[1 mark]***. This reduces the concentration gradient of oxygen in the alveoli ***[1 mark]***, which reduces the rate at which oxygen diffuses into the blood ***[1 mark]***. This may result in shortness of breath, because the person with bronchitis may feel like they have to breathe faster/harder as the body tries to make up for a lack of oxygen reaching the cells ***[1 mark]***.

b) The alveoli are surrounded by a capillary network, so blood constantly takes oxygen away from the alveoli and brings carbon dioxide ***[1 mark]***. The alveoli are ventilated, so the air in them is constantly being replaced, with oxygen being brought in and carbon dioxide being removed ***[1 mark]***.

c) The breakdown of the alveoli walls would mean the loss of elastic fibres from the alveoli ***[1 mark]***. This would make it harder for the lungs to recoil and expel air ***[1 mark]***.

Module 3 : Section 2 — Transport in Animals

Pages 47-49: Transport in Animals — 1

1 A ***[1 mark]***

The ECG shows a very irregular heartbeat, which is fibrillation. Tachycardia is when the heartbeat is too fast (over 100 beats per minute), an ectopic heartbeat is when there's an 'extra' heartbeat (an early contraction of the atria or ventricles) and bradycardia is when the heartbeat is too slow (under 60 beats per minute).

2 C ***[1 mark]***

Statement 1 is incorrect — at low partial pressures of oxygen, haemoglobin unloads oxygen (oxyhaemoglobin dissociates).

3 D ***[1 mark]***

4 a) It conducts the waves of electrical activity, after a short delay, from the SAN to the bundle of His ***[1 mark]***.

b) The layer of collagen prevents the wave of electricity passing straight from the atria into the ventricles ***[1 mark]***. This makes sure that the ventricles only respond to signals from the AVN, transferred via the bundle of His/Purkyne tissue ***[1 mark]***.

5 a) They prevent the backflow of blood ***[1 mark]*** into the atria when the ventricles contract ***[1 mark]***.

b) They open when pressure is greater below the valve / in the ventricle than in the artery ***[1 mark]***.

c) i)

	Standing up	Lying down
Mean heart rate (bpm)	74	57
Mean cardiac output (cm^3 min^{-1})	4700	4700
Mean stroke volume (cm^3)	**63.5** (3 s.f.)	**82.5** (3 s.f.)

[2 marks for both correct answers, otherwise 1 mark for evidence of 4700 ÷ 74 or 4700 ÷ 57]

Cardiac output = heart rate × stroke volume, so stroke volume = cardiac output ÷ heart rate. So when standing up, stroke volume is 4700 ÷ 74 = 63.5, and when lying down it's 4700 ÷ 57 = 82.5.

ii) E.g. it gives the heart rate time to stabilise, as the act of changing position could cause it to increase ***[1 mark]***.

iii) E.g. heart rate is lower when lying down as blood does not have to be pumped above the level of the heart / against gravity ***[1 mark]***.

The heart has to work harder when a person is standing up, because blood has to flow against gravity. When you're lying down, the force of gravity is evenly distributed across the body.

iv) Taking multiple measurements and calculating the mean reduces the effect of random error, so makes the results more precise ***[1 mark]***.

Pages 50-53: Transport in Animals — 2

1 a) i) X = aorta ***[1 mark]***
Y = pulmonary artery ***[1 mark]***

ii) E.g. the artery walls would have felt much thicker than the vein walls ***[1 mark]***.

b)

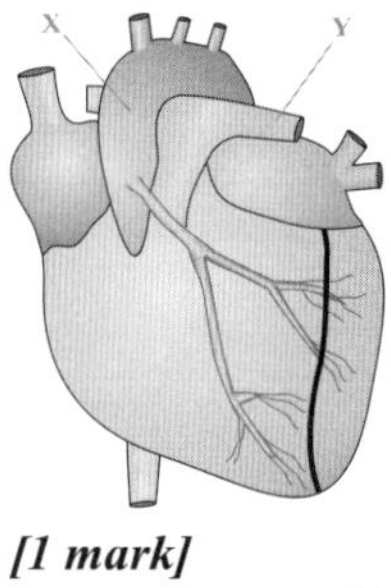

[1 mark]

c) coronary artery ***[1 mark]***

d) Any one from: e.g. use clear, continuous lines/no overlaps in lines ***[1 mark]*** / no shading ***[1 mark]*** / draw different components in proportion ***[1 mark]*** / include a scale ***[1 mark]*** / include relevant labels ***[1 mark]***.

2 a) i) E.g. there is a build-up of tissue fluid because excess fluid is not being returned to the blood via the lymphatic system ***[1 mark]***.

ii) Any two from: e.g. lymph has no red blood cells ***[1 mark]***. / Lymph has no platelets ***[1 mark]***. / Lymph has fewer proteins ***[1 mark]***.

b) i) Fluid leaves the capillaries at the arteriole end of the capillary bed ***[1 mark]***, reducing the hydrostatic pressure at the venule end of the capillary bed ***[1 mark]***.

ii) Oncotic pressure is generated by plasma proteins in the capillaries ***[1 mark]***. The plasma proteins remain in the capillaries because they're too large to fit through the capillary walls, so oncotic pressure doesn't change along the capillary bed ***[1 mark]***.

c) E.g. capillaries have walls that are only one cell thick, which shortens the diffusion pathway ***[1 mark]***.

3 a) The partial pressure of oxygen is lower at point A than at point B ***[1 mark]***. The lower the partial pressure of oxygen, the lower the affinity haemoglobin has for oxygen, and so the percentage saturation of haemoglobin will be lower at point A than point B ***[1 mark]***.

b) i) The respiration rate increases during exercise, which increases the partial pressure of carbon dioxide in the blood ***[1 mark]***. Higher partial pressures of carbon dioxide increase the rate of oxygen unloading (the Bohr effect) and the saturation of blood with oxygen is lower for a given partial pressure of O_2 ***[1 mark]***.

ii) Carbonic anhydrase catalyses the reaction between carbon dioxide and water in red blood cells to form carbonic acid ***[1 mark]***. Carbonic acid then dissociates to form HCO_3^- and H^+ ions ***[1 mark]***. The increase in H^+ ions causes haemoglobin to unload its oxygen ***[1 mark]***.

4 a) 60 ÷ 0.55 = 109.0909...
109 beats per minute ***[1 mark]***

b) 0.13 seconds ***[1 mark]***

The left atrium contracts before the left ventricle in the cardiac cycle, so you need to find a point on the graph where the pressure of the atrium increases before the pressure of the ventricle increases.

c) How to grade your answer:
Level 0: There is no relevant information. ***[No marks]***
Level 1: The answer covers one or two of the points A to D. The answer has no clear structure. The information given is basic and lacking in detail. It may not all be relevant. ***[1 to 2 marks]***
Level 2: The answer covers at least three of the points A to D. The answer has some structure. Most of the information given is relevant and there is some detail involved. ***[3 to 4 marks]***
Level 3: The answer covers all four of the points A to D in detail. The answer has a clear and logical structure. The information given is relevant and detailed. ***[5 to 6 marks]***

Indicative scientific content may include:

Point A
Pressure in the left ventricle exceeds pressure in the left atrium, because the left ventricle is contracting and the left atrium is relaxing. This causes the atrioventricular valve/the valve between the left atrium and left ventricle to close, preventing the backflow of blood into the left atrium.

Point B
Pressure increases in the left ventricle to above that of the aorta, which forces the semi-lunar valve/the valve between the left ventricle and aorta open.

Point C
The left ventricular pressure falls below that of the aorta, because blood has moved into the aorta from the ventricle and the left ventricle is relaxing. As a result, the semi-lunar valve/the valve between the left ventricle and the aorta closes.

Point D
Pressure has been increasing in the left atrium as blood has been returning to the atrium from the body. As the atrial pressure exceeds ventricular pressure, the atrioventricular valve/valve between the left atrium and left ventricle opens, allowing blood to flow into the left ventricle.

d) The left ventricle has a higher maximum pressure than the left atrium because it has a thicker muscle wall and so is able to generate more force when it contracts ***[1 mark]***.

e) The wall of the aorta is thick and muscular / the wall of the aorta contains elastic tissue to stretch and recoil / the inner lining of the aorta is folded so it can stretch ***[1 mark]***, which helps to maintain the high pressure of the blood coming out of the left ventricle ***[1 mark]***.

Module 3 : Section 3 — Transport in Plants

Pages 54-57: Transport in Plants

1 B ***[1 mark]***

2 D ***[1 mark]***

Water travelling via the symplast pathway travels through plasmodesmata (channels in the cells walls). Water travelling via the apoplast pathway travels through the cell walls themselves.

3 C ***[1 mark]***

Statement 1 is incorrect. Sunken stomata and a thick waxy epidermis are xerophytic adaptations — they help a plant to conserve water. Hydrophytes live in water, so they don't need help conserving it. Instead they need adaptations to help them get enough oxygen.

4 a) Water on the leaves would reduce the water potential gradient between inside the leaf and outside ***[1 mark]***, reducing water loss/transpiration ***[1 mark]***.

b) E.g. cut the shoot underwater / set the apparatus up underwater / check the apparatus is watertight ***[1 mark]***.

c) E.g. it is a measurement of the rate of water uptake only, which is not exactly the same as transpiration rate/the rate of water loss from the leaves ***[1 mark]***. This is because some of the water taken up by the plant is used for support/turgidity/photosynthesis ***[1 mark]***.

d) At higher temperatures, water molecules have more kinetic energy so they evaporate more quickly from the cells inside the leaf ***[1 mark]***. This increases the water potential/concentration gradient between the inside and outside of the leaf ***[1 mark]***, so water diffuses out of the leaf faster and the rate of transpiration is faster ***[1 mark]***.

5 a) Light intensity is higher at 12:00, so more stomata are open ***[1 mark]***, which increases the transpiration rate ***[1 mark]***. This draws water molecules up the xylem at a quicker rate, due to cohesion and tension ***[1 mark]***.

At 00:00, it would be dark, whereas there is sunlight at 12:00.

b) i) To prevent the plant tissue from drying out ***[1 mark]***.

ii) E.g. dyeing the tissue with a stain/named stain ***[1 mark]***.

iii) E.g. in a root, the xylem is arranged in the centre, whereas in a stem, the xylem is near the outside. / In a root, the xylem is grouped together in a cross shape, whereas in a stem, it is arranged in individual bundles that form a ring around the stem. ***[1 mark]***

6 a) Translocation requires energy/ATP ***[1 mark]***. If metabolism stops, respiration cannot occur so ATP is not produced / energy is not released ***[1 mark]***.

b) E.g. at point A, the water potential is being lowered by solutes being actively loaded from the companion cell into the sieve tube ***[1 mark]***. This causes water to enter from the xylem and companion cell, and raises the pressure ***[1 mark]***. At point B, the pressure is lower because water is leaving the sieve tube by osmosis ***[1 mark]***. This is occurring because solutes are being removed from the sieve tube and so the water potential is higher ***[1 mark]***.

Module 4 : Section 1 — Disease and the Immune System

Pages 58-60: Disease and the Immune System — 1

1 B ***[1 mark]***

2 D ***[1 mark]***

3 a) i) It agglutinates pathogens/causes pathogens to clump together ***[1 mark]***.

ii) B lymphocytes / plasma cells / memory B cells ***[1 mark]***

iii) Any two from: e.g. its variable region is complementary to a particular antigen, allowing it to bind to that antigen. / The hinge region allows flexibility when the antibody binds to the antigen. / The constant region allows the antibody to bind to receptors on immune system cells. ***[2 marks]***

b) E.g. it allows the antibody to bind to more antigens at once ***[1 mark]***, so there is a greater chance of agglutination occurring / more pathogens can be phagocytosed at once ***[1 mark]***.

c) $2000 \times 60 \times 60 = 7200000$ ***[1 mark]***
$= \mathbf{7.2 \times 10^6}$ ***[1 mark]***

There are 60 seconds in a minute and 60 minutes in an hour. So, to work out how many molecules could be produced in an hour, multiply the number that can be produced in one second by 60, and then by 60 again.

4 a) Indirect transmission, because the disease is transmitted via an intermediate (contaminated water) ***[1 mark]***.

b) i) E.g. there is a positive correlation between the number of deaths from the disease and monthly rainfall in this country ***[1 mark]***.

ii) E.g. higher rainfall may increase flooding, making it more likely for infected human faeces to get into sources of drinking water ***[1 mark]***. This could increase the transmission of the pathogen between people, resulting in a greater number of deaths from the disease ***[1 mark]***.

c) E.g. each strain of the pathogen will have different, specific antigens on its surface ***[1 mark]***. A specific antigen must be bound by an antibody with a complementary shape ***[1 mark]***, so the vaccine will need to contain different antibodies for the antigens on each strain ***[1 mark]***.

Pages 61-64: Disease and the Immune System — 2

1 a) E.g. it would reduce damage caused by caterpillars to the outer surfaces of the plant, which would reduce the number of sites through which pathogens could enter the plant. / If the caterpillars are vectors for a disease, it could reduce the likelihood of them transmitting the pathogen they carry to the corn plant. ***[1 mark]***

b) E.g. its cells are surrounded by cell walls, which would form a physical barrier against entry by the pathogen. / It may produce callose depositions (between its cell walls and its cell membranes) that make it harder for the pathogen to enter its cells. ***[1 mark]***

2 a) i) They move in response to signals from cytokines ***[1 mark]***, which are released by damaged cells/tissue at the site of the wound ***[1 mark]***.

ii) E.g. it means that any pathogens entering the body through the wound can be phagocytosed/destroyed before they are able to spread ***[1 mark]***.

b) Any three from: e.g. the phagocyte recognises the antigens on the pathogen as foreign ***[1 mark]***. The cytoplasm of the phagocyte moves around the pathogen, engulfing it ***[1 mark]***. This results in the formation of a phagosome, containing the pathogen ***[1 mark]***. A lysosome fuses with the phagosome, releasing digestive enzymes to break down the pathogen / the enzymes in a lysosome break down the pathogen ***[1 mark]***. The phagocyte then presents the pathogen's antigens on its surface ***[1 mark]***.

c) E.g. phagocytosis may be less effective ***[1 mark]*** as opsonins/IgG bind(s) to foreign antigens to aid phagocytosis / allow the phagocyte to get closer to the pathogen in order to engulf it ***[1 mark]***.

3 a) i) E.g. the antibody concentration remains low for several days after exposure to the antigen, while clonal selection and clonal expansion of B lymphocytes occurs ***[1 mark]***. Antibody concentration increases as plasma cells (produced as a result of clonal expansion) start to rapidly produce antibodies that are complementary to the antigen ***[1 mark]***. Antibody concentration reaches a gradual peak and then falls as plasma cells die off ***[1 mark]***. It doesn't fall back to pre-exposure levels because memory B lymphocytes (produced by clonal expansion) remain in the body ***[1 mark]***.

ii) E.g. the secondary immune response is faster/stronger than the primary immune response / the concentration of the complementary antibody increases rapidly ***[1 mark]*** allowing the pathogen to be destroyed before the person experiences any symptoms ***[1 mark]***.

b) Anti-toxins bind to complementary toxins that are produced by the bacteria ***[1 mark]***. This neutralises the toxins and prevents them from causing muscle spasms ***[1 mark]***. However, anti-toxins don't bind to the antigens on the *C. tetani* bacteria themselves, so the bacteria would remain in the body if the patient was only treated with anti-toxins ***[1 mark]***.

4 a) i) E.g. so that the woman produces antibodies against the bacteria, which would be transferred to her baby (via the placenta) before it is born ***[1 mark]***.

ii) Any four from: e.g. the vaccine contains GBS antigens ***[1 mark]***. Neutrophils/phagocytes/antigen-presenting cells present these antigens to T helper cells ***[1 mark]***. When the antigens bind to receptors on T helper cell membranes, the T helper cells are activated ***[1 mark]***. The activated T helper cells then release interleukins/cytokines/chemical messengers to activate B lymphocytes ***[1 mark]***. Both B lymphocytes and T helper cells undergo clonal expansion to produce memory cells ***[1 mark]***.

b) Any two from: e.g. the immunity the baby receives from its mother is immediate, whereas immunity from a vaccine takes time to develop. / A vaccine involves exposing the baby to an antigen, whereas breastfeeding provides immunity without needing to expose the baby to an antigen. / The baby does not produce memory cells as a result of breastfeeding, but memory cells are produced as a result of a vaccine. / The immunity the baby gets from breastfeeding only lasts for a short time, whereas the immunity from a vaccine lasts much longer. ***[2 marks]***

c) i) E.g. the GBS bacteria damage the protective layers around the brain causing the tissue to release messenger molecules ***[1 mark]***. The messenger molecules increase the permeability of nearby blood vessels/cause vasodilation ***[1 mark]***. This causes fluid to leak into the area, resulting in swelling/this increases blood flow to the area, making it hot ***[1 mark]***

ii) E.g. the bacteria that cause meningitis may become resistant to the antibiotics used to kill them, making it harder to treat the disease ***[1 mark]***.

5 a) How to grade your answer:
Level 0: There is no relevant information. ***[No marks]***
Level 1: The answer briefly describes/explains something that would happen if a person with blood type A received a transfusion of blood type B. The answer has no clear structure. The information given is basic and lacking in detail. It may not all be relevant. ***[1 to 2 marks]***
Level 2: The answer describes and explains most of what would happen if a person with blood type A received a transfusion of blood type B. The answer has some structure. Most of the information given is relevant and there is some detail involved. ***[3 to 4 marks]***
Level 3: The answer fully describes and explains what would happen if a person with blood type A received a transfusion of blood type B. The answer has a clear and logical structure. The information given is relevant and detailed. ***[5 to 6 marks]***
Indicative scientific content may include:
The antigens on type B red blood cells are different/a different shape to the antigens on type A red blood cells.
As a result, the type A person's phagocytes will not recognise the antigens on the type B red blood cells / will recognise that the type B antigens are foreign.
This will cause a phagocyte that encounters a type B red blood cell to engulf the cell.
The type B red blood cell is broken down by the phagocyte and its antigens are presented on the phagocyte's surface.
T lymphocytes with complementary receptors bind to the presented antigens. This causes the T lymphocytes to undergo clonal selection.
The selected T lymphocytes then undergo clonal expansion.
Some of these T lymphocytes become T helper cells.
B lymphocytes with complementary receptors also bind to the presented antigens.
This binding, along with substances released by T helper cells, causes the B lymphocytes to undergo clonal selection.
Selected B lymphocytes divide/undergo clonal expansion to produce plasma cells.
The plasma cells secrete antibodies against the antigens on the type B red blood cells.
The antibodies bind to the type B antigens causing the type B red blood cells to clump together/agglutinate.
The agglutinated type B red blood cells are destroyed by phagocytosis.

b) Blood type O has no antigens, so no immune response will be triggered ***[1 mark]***.

c) E.g. agglutinins for type A and type B antigens could be added to a sample of the person's blood ***[1 mark]***. If no agglutination/clumping is observed, then the person must have type O blood ***[1 mark]***.

Module 4 : Section 2 — Biodiversity

Pages 65-68: Biodiversity

1 B ***[1 mark]***

2 B ***[1 mark]***

Statements 1 and 2 are both correct. Statement 3 is incorrect because marine conservation zones involve protecting species in their natural habitat, whereas ex situ conservation means removing organisms from their habitat and placing them in a new location.

3 C ***[1 mark]***

Statement 1 might be true but you can't tell from this data alone because Simpson's Index takes into account both species richness and species evenness. Statement 2 is definitely false because the closer the value of Simpson's Index to O, the less diverse the habitat — so Habitat B actually has the lowest biodiversity. The closer the value is to 1, the more diverse the habitat, so Habitat C is more diverse than Habitat D and statement 3 is the only statement that you can say is definitely true.

4 a) i)

Species	Number of individual plants counted in different quadrats					Mean number counted
Rapeseed	24	46	32	28	32	**32.4**
Common sunflower	1	0	2	1	1	**1**
Common poppy	8	12	6	10	8	**8.8**
Creeping thistle	13	14	7	15	13	**12.4**

[1 mark]

ii) $N = 32.4 + 1 + 8.8 + 12.4 = 54.6$

$$D = 1 - \left(\left(\frac{32.4}{54.6}\right)^2 + \left(\frac{1}{54.6}\right)^2 + \left(\frac{8.8}{54.6}\right)^2 + \left(\frac{12.4}{54.6}\right)^2\right)$$

$D =$ **0.570** (to 3 s.f.)

[2 marks for the correct answer, otherwise 1 mark for correct working. Allow full marks if incorrect answers to 4 a) i) used correctly here.]

b) E.g. the field could have been divided into a grid and a random number generator could have been used to select coordinates at which to place the quadrats ***[1 mark]***.

c) The scientists could have used a Student's t-test ***[1 mark]*** as this would have allowed them to compare two mean values ***[1 mark]***.

d) E.g. animal grazing prevents some plants from growing, reducing plant diversity ***[1 mark]***. This would mean there are fewer habitats and food resources to support other organisms, further reducing species diversity ***[1 mark]***.

5 a) Because an index of diversity takes into account both species richness and species evenness ***[1 mark]***, which means that it takes into account species that are only present in small numbers, which species richness does not ***[1 mark]***.

b) Any three from: e.g. there were fewer ladybird species on the conventional farm than on the organic farm. / The standard deviation bars do not overlap, so the difference was significant/the standard deviation bars are short, showing that the data is precise. / However, there is no indication of how many samples were taken, so the data may not be representative of all organic/conventional farms. / There may be factors other than the way in which the fields were farmed, which influenced the number of ladybird species present. / The scientists' conclusion is correct for this data, but further investigation is needed to ensure that these results are valid ***[1 mark for each correct answer]***.

c) i) E.g. habitat diversity is lost because large areas of land need to be cleared to make way for monoculture. / Species diversity is lost because only one type of crop is grown and naturally occurring plants and animals are removed using pesticides. ***[1 mark]***

ii) E.g. continuous monoculture causes soil depletion ***[1 mark]***. This means a lot of money needs to be spent on fertilisers to artificially replace soil nutrients / yields can decrease in the long run, reducing the profit that can be made from the land ***[1 mark]***.

6 a) i) E.g. the proportion of polymorphic gene loci indicates the genetic diversity of the population / genetic diversity is likely to be low in a population of animals bred in captivity ***[1 mark]***. Monitoring it will help the breeding programme to be managed so that genetic diversity is maximised ***[1 mark]***.

ii) number of polymorphic gene loci = 165 – 132 = 33
proportion of polymorphic gene loci =
number of polymorphic ÷ total number of loci =
33 ÷ 165 = **0.2**
[2 marks for correct answer, otherwise 1 mark for using the correct formula]

Polymorphism is when a locus has more than one allele.

b) Any one from: e.g. without beavers the ecosystem would be dramatically different. / Many organisms depend on the habitats created and maintained by beavers. ***[1 mark]***

c) Any one from: e.g. habitats near rivers may have been cleared to make way for more housing/road developments/ farmland. / River habitats may have been polluted by increased use of pesticides/fertilisers. / River habitats may have been over-exploited/over-fished to meet an increasing demand for food. ***[1 mark]***

Human population growth has also contributed to climate change, so you could have written about that too.

d) E.g. the Convention on International Trade in Endangered Species (CITES) regulates the international trade in wild animal and plant specimens ***[1 mark]***. The countries that signed it agreed to make it illegal to kill endangered species/ban the trade of products made from endangered species ***[1 mark]***.

You could have written about another agreement here instead, e.g. the Rio Convention on Biological Diversity (CBD) or the Countryside Stewardship Scheme (CSS).

Module 4 : Section 3 — Classification and Evolution

Pages 69-70: Classification and Evolution — 1

1 B ***[1 mark]***

2 D ***[1 mark]***

Statement 2 is false because the most recent common ancestor of the blind snakes, pythons and boas is the same as the most recent common ancestor of the blind snakes and vipers, so blind snakes are equally related to either group. Statement 3 is also false, because elapids, Atractaspis *and colubrids diverged at the same point.*

3 a) four ***[1 mark]***

The genera shown on the phylogenetic tree are Panthera, Meles, Lutra *and* Canis. *You know that these are the genera because the binomial name for a species is made up of two parts, e.g.* Panthera leo, *and the first part is always the genus.*

b) *Canis aureus* and *Canis lupus* because they both belong to the same genus ***[1 mark]***.

c) family ***[1 mark]***

d) Molecular evidence showed that the kingdom of Prokaryotae could be split into the domains of Archaea and Bacteria ***[1 mark]***, e.g. the enzyme RNA polymerase was found to be different in Bacteria and Archaea ***[1 mark]***.

Pages 71-73: Classification and Evolution — 2

1 a) Concentrating their urine allows them to conserve water, which means they are less likely to become dehydrated in their hot desert habitat ***[1 mark]***.

b) Some mice had an allele/alleles for golden-brown fur ***[1 mark]***. This would have been beneficial because it would have been harder for predators to see these mice ***[1 mark]***, so they would have been more likely to survive long enough to reproduce and pass on their beneficial alleles to the next generation ***[1 mark]***. After many generations, the frequency of the allele/alleles for golden-brown fur would have increased in the population ***[1 mark]***.

c) Every species has a Latin/binomial name ***[1 mark]***, which avoids confusion, as there might be other common names than 'golden spiny mice'/because binomial names are internationally recognised ***[1 mark]***.

d) i) The variation in mass could have been caused by genetic factors, as these individuals could have inherited alleles that caused them to have a greater mass ***[1 mark]***. It could also have been caused by environmental factors, such as these mice having more food available to them, causing them to become heavier ***[1 mark]***.

ii) $$\overline{x} = \frac{43.9 + 39.9 + 42.6 + 43.1 + 37.4}{5} = 41.38$$

$$\sum(x-\overline{x})^2 = (43.9 - 41.38)^2 + (39.9 - 41.38)^2 + (42.6 - 41.38)^2 + (43.1 - 41.38)^2 + (37.4 - 41.38)^2 = 28.828$$

$$s = \sqrt{\frac{28.828}{5-1}} = 2.68458... = \mathbf{2.68\ (3\ s.f.)}$$

[2 marks for correct answer, otherwise 1 mark for correct working]

iii) By carrying out a Student's t-test ***[1 mark]***.

Student's t-tests are for comparing the means of two sets of data.

2 a) E.g. Wallace came up with the idea of natural selection as a mechanism for evolution at the same time as Darwin ***[1 mark]***, and provided evidence to support it, e.g. he wrote about advantageous adaptations that had evolved by natural selection, such as warning colours to deter predators ***[1 mark]***.

b) The *Archaeopteryx* fossils show that there could have been evolution from land dinosaurs into birds ***[1 mark]***, because the *Archaeopteryx* had some of the features of the land dinosaurs they may have evolved from, but also some of the features of birds, which they could have evolved into over time ***[1 mark]***.

c) E.g. scientists can compare molecules (e.g. proteins) in different organisms to find similarities and differences ***[1 mark]***. Those organisms that have diverged away from each other more recently and are more closely related will have more similar molecules ***[1 mark]***.

3 How to grade your answer:

Level 0: There is no relevant information. ***[No marks]***

Level 1: The answer covers how refuge areas reduce the spread of Bt corn resistance or the implications of reducing the spread of Bt resistance for human populations. The answer has no clear structure. The information given is basic and lacking in detail. It may not all be relevant. ***[1 to 2 marks]***

Level 2: The answer covers how refuge areas reduce the spread of Bt corn resistance and some of the implications of reducing the spread of Bt resistance for human populations. The answer has some structure. Most of the information given is relevant and there is some detail involved. ***[3 to 4 marks]***

Level 3: The answer fully covers the how refuge areas reduce the spread of Bt corn resistance and the implications of reducing the spread of Bt resistance for human populations.
The answer has a clear and logical structure.
The information given is relevant and detailed. ***[5 to 6 marks]***

Indicative scientific content may include:

How refuge areas reduce the spread of Bt corn resistance

Without refuge areas, the only European corn borers that survive long enough to reproduce would be resistant ones, so their offspring would inherit the alleles for resistance and the whole population of European corn borers would continue to be resistant.

In refuge areas, European corn borers may be either resistant or non-resistant, because there is no pesticide acting as a selection pressure. This means that Bt-resistant European corn borers may mate with non-resistant individuals from refuge areas to produce non-resistant offspring. This would reduce the number of resistant corn borers in the population.

Implications for human populations

By reducing the spread of Bt corn resistance, the destruction of Bt corn crops by European corn borers can be kept to a minimum. This would help to feed the growing human population, and prevent farmers from losing large amounts of their income to pest damage.

Reducing the spread of Bt corn resistance would also mean that farmers do not need to turn to alternative pesticides, which may be less effective and less specific, killing insects that are beneficial for the crop (such as pollinators).

Finding an adequate replacement pesticide might require the research and production of new pesticides, which costs time and money.

Mixed Questions

Pages 74-76: Mixed Questions — 1

1 C ***[1 mark]***

Fish have a closed, single circulatory system. Both sides of the heart pump blood to the gills and to the rest of the body. Veins take blood back to the heart.

2 B ***[1 mark]***

A pyrimidine base contains a single carbon-nitrogen ring, so it must be B. Molecule D contains two carbon-nitrogen rings joined together — it's a purine base. Molecule A is α-glucose, and molecule C is an amino acid.

3 A ***[1 mark]***

You can tell that this image was taken using a scanning electron microscope because you can see more detail than you could with a light or laser scanning confocal microscope, and the image is 3D (transmission electron microscopes only produce 2D images).

4 a) Carbohydrate / disaccharide ***[1 mark]***

b) i) E.g. the reaction/breakdown of lactose is a hydrolysis reaction ***[1 mark]***, which requires the addition of a water molecule ***[1 mark]***.

ii) Any two from: e.g. the temperature of the solutions. / The concentration of the lactose and lactase solutions. / The type of glucose test strip used. / The person observing the colour change of each glucose test strip. ***[2 marks available — 1 mark for each reasonable suggestion]***

For this question there are other correct answers that aren't listed here. For an answer to be valid, the variable needs to have an impact on the results, and be something that's possible to control in an experiment.

c) i) E.g. that lactase is only active below pH 8 / lactase is inactive at pH 8 or above ***[1 mark]***. Lactase breaks lactose down into glucose and galactose, therefore, if glucose is present/the strip is green, it indicates lactase is active / if glucose is absent/the strip is yellow, it indicates that lactase is inactive ***[1 mark]***.

ii) Above and below its optimum pH, every enzyme is affected by the H^+ and OH^- ions found in acids and alkalis ***[1 mark]***. These ions interfere with the ionic and hydrogen bonds that hold the enzyme's tertiary structure together ***[1 mark]***. This changes the shape of the enzyme's active site ***[1 mark]***.

iii) E.g. increase the number of reactions carried out between pH 7 and 8 ***[1 mark]*** because it is between these pH values that lactase activity stops ***[1 mark]***.

Accurate results need to be close to the true answer, so increasing the number of reactions carried out between pH 7 and 8 would let the student narrow down where lactase activity stops. E.g. the investigation could show that the activity stops between pH 7.2 and 7.3 rather than between pH 7 and 8.

d) If the mutation alters the sequence of amino acids/primary structure of lactase ***[1 mark]***, it could cause the shape of lactase's active site to change ***[1 mark]***. This may mean that lactose is no longer able to fit into/bind to the active site, so lactase is prevented from catalysing its breakdown ***[1 mark]***.

e) i) E.g. phagocytes may identify the milk proteins as foreign antigens ***[1 mark]***. They would then move their cytoplasm around the proteins, engulfing them ***[1 mark]***. The milk proteins would then be broken down in the phagocytes by digestive enzymes ***[1 mark]***. The phagocytes would then present the fragments on their surfaces to activate other immune cells ***[1 mark]***.

ii) The secondary immune response to a foreign antigen is quicker than the primary immune response ***[1 mark]***. This is because clonal selection happens faster and memory T and B lymphocytes are activated ***[1 mark]***.

Pages 77-79: Mixed Questions — 2

1 a) i) spindle fibre ***[1 mark]***

ii) Cell B because the diagram shows the homologous pairs being separated, not the sister chromatids ***[1 mark]***.

b) 20 ***[1 mark]***

c) i) Meiosis because the chromosome number changes from diploid/2n in the sporophyte to haploid/n in the spores ***[1 mark]***.

ii) When a sperm fertilises an egg, a diploid/2n fertilised gametophyte is formed ***[1 mark]***. This must mean that both the egg and sperm are haploid/n ***[1 mark]***. The spores are also haploid/n and to maintain a haploid number of chromosomes between the spores and gametophytes, stages 1 and 2 must represent mitosis ***[1 mark]***.

d) i) E.g. there may be lots of variation in habitat in each area ***[1 mark]*** and she may want to make sure that the different habitats are sampled in proportion ***[1 mark]***.

ii) E.g. repeat the sampling at least three times and calculate a mean for each area ***[1 mark]***.

2 a) Cl^- ions move from the epithelial cells lining the ileum, through the open Cl^- channel proteins and into the ileum lumen ***[1 mark]***. The build up of Cl^- ions lowers the water potential of the lumen ***[1 mark]***. This causes water to move by osmosis out of the epithelial cells into the lumen ***[1 mark]***. This lowers the water potential of the epithelial cells, causing water to move out of the blood into the cells ***[1 mark]***.

b) i) E.g. *V. cholerae* will have a cell wall, the ileum epithelial cell will not. / *V. cholerae* will not have a nucleus/ membrane-bound organelles, the ileum epithelial cell will. / *V. cholerae* will have smaller ribosomes than the ileum epithelial cell ***[1 mark]***.

You're being asked to compare the structure of a prokaryotic cell (V. cholerae) to that of a eukaryotic animal cell (the ileum epithelial cell) here.

ii) Any four from: e.g. diffusion of nutrients across the outer surface alone would be too slow to meet a human's needs ***[1 mark]***. This is because it would take too long for substances to travel from the outer surface to the cells deep within the body ***[1 mark]***. Not enough substances would be exchanged across the outer surface, which is small relative to the volume of a human ***[1 mark]***. Humans have a relatively high metabolic rate, so diffusion across the outer surface would be too slow to keep up with metabolic demand ***[1 mark]***. Therefore humans need a specialised exchange surface to improve the efficiency of nutrient absorption ***[1 mark]***. ***[Maximum of 4 marks available]***

c) Any two from: e.g. drink only boiled/treated water. / Cook food thoroughly. / Wash hands often with soap and water. / Improve sewage/water treatment. / Keep sewage outlets away from sources of drinking water. ***[2 marks]***.

d) Severely ill patients may not be able to overcome a cholera infection without antibiotics/through their immune response alone, unlike less severely ill patients ***[1 mark]***. By restricting the use of antibiotics, the chance of *V. cholerae* developing antibiotic resistance will be reduced, making it easier to treat severe cholera infections in future ***[1 mark]***.